VOLUME ONE HUNDRED AND TWENTY EIGHT

Progress in Molecular Biology and Translational Science

Epigenetics and Neuroplasticity—Evidence and Debate

VOLUME ONE HUNDRED AND TWENTY EIGHT

PROGRESS IN MOLECULAR BIOLOGY AND TRANSLATIONAL SCIENCE

Epigenetics and Neuroplasticity—Evidence and Debate

Edited by

SCHAHRAM AKBARIAN

Department of Psychiatry, and Department of Neuroscience, Icahn School of Medicine at Mount Sinai, New York, USA

FARAH LUBIN

Department of Neurobiology at the University of Alabama at Birmingham, Evelyn F. McKnight Brain Institute, Birmingham, Alabama, USA

ELSEVIER

AMSTERDAM • BOSTON • HEIDELBERG • LONDON
NEW YORK • OXFORD • PARIS • SAN DIEGO
SAN FRANCISCO • SINGAPORE • SYDNEY • TOKYO
Academic Press is an imprint of Elsevier

Academic Press is an imprint of Elsevier
225 Wyman Street, Waltham, MA 02451, USA
525 B Street, Suite 1800, San Diego, CA 92101-4495, USA
32 Jamestown Road, London NW1 7BY, UK
The Boulevard, Langford Lane, Kidlington, Oxford OX5 1GB, UK

First edition 2014

Notices

Knowledge and best practice in this field are constantly changing. As new research and experience broaden our understanding, changes in research methods, professional practices, or medical treatment may become necessary.

Practitioners and researchers must always rely on their own experience and knowledge in evaluating and using any information, methods, compounds, or experiments described herein. In using such information or methods they should be mindful of their own safety and the safety of others, including parties for whom they have a professional responsibility.

To the fullest extent of the law, neither the Publisher nor the authors, contributors, or editors, assume any liability for any injury and/or damage to persons or property as a matter of products liability, negligence or otherwise, or from any use or operation of any methods, products, instructions, or ideas contained in the material herein.

ISBN: 978-0-12-800977-2
ISSN: 1877-1173

For information on all Academic Press publications visit our website at store.elsevier.com

CONTENTS

CONTRIBUTORS

Schahram Akbarian
Department of Psychiatry, and Department of Neuroscience, Icahn School of Medicine at Mount Sinai, New York, USA

Angel Barco
Instituto de Neurociencias, Universidad Miguel Hernández-Consejo Superior de Investigaciones Científicas, Alicante, Spain

Elisabeth B. Binder
Department of Translational Research in Psychiatry, Max Planck Institute of Psychiatry, Munich, Germany, and Department of Psychiatry and Behavioral Sciences, Emory University School of Medicine, Atlanta, Georgia, USA

Erbo Dong
The Psychiatric Institute, Department of Psychiatry, College of Medicine, University of Illinois at Chicago, Chicago, Illinois, USA

Janina Galler
Judge Baker Children's Center, Harvard Medical School, Boston, Massachusetts, USA

Dennis R. Grayson
The Psychiatric Institute, Department of Psychiatry, College of Medicine, University of Illinois at Chicago, Chicago, Illinois, USA

Alessandro Guidotti
The Psychiatric Institute, Department of Psychiatry, College of Medicine, University of Illinois at Chicago, Chicago, Illinois, USA

Leah N. Hitchcock
Department of Behavioral Neuroscience, Oregon Health & Science University, Portland, Oregon, USA

Timothy J. Jarome
Department of Neurobiology, University of Alabama at Birmingham, Birmingham, AL, USA

K. Matthew Lattal
Department of Behavioral Neuroscience, Oregon Health & Science University, Portland, Oregon, USA

Robert H. Lipsky
Inova Neurosciences Institute, Inova Health System, Falls Church, and Department of Molecular Neuroscience, The Krasnow Institute for Advanced Study, George Mason University, Fairfax, Virginia, USA

Sermsak Lolak
Department of Psychiatry, The George Washington University School of Medicine and Health Sciences, Washington, District of Columbia, USA

Jose P. Lopez-Atalaya
Instituto de Neurociencias, Universidad Miguel Hernández-Consejo Superior de Investigaciones Científicas, Alicante, Spain

Farah D. Lubin
Department of Neurobiology, University of Alabama at Birmingham, Birmingham, AL, USA

Amanda Mitchell
Department of Psychiatry, and Department of Neuroscience, Icahn School of Medicine at Mount Sinai, New York, USA

Cyril Peter
Department of Psychiatry, and Department of Neuroscience, Icahn School of Medicine at Mount Sinai, New York, USA

Nadine Provençal
Department of Translational Research in Psychiatry, Max Planck Institute of Psychiatry, Munich, Germany

Danielle Galler Rabinowitz
Judge Baker Children's Center, Harvard Medical School, Boston, Massachusetts, USA

Carina Rampp
Department of Translational Research in Psychiatry, Max Planck Institute of Psychiatry, Munich, Germany

Panos Roussos
Department of Psychiatry; Department of Neuroscience, and Department of Genetics and Genomic Sciences, Icahn School of Medicine at Mount Sinai, New York, USA

Pim Suwannarat
Department of Genetics, Mid-Atlantic Permanente Medical Group, Rockville, Maryland, USA

Jasmyne S. Thomas
Department of Neurobiology, University of Alabama at Birmingham, Birmingham, AL, USA

Nadejda Tsankova
Department of Neuroscience, and Department of Pathology, Icahn School of Medicine at Mount Sinai, New York, USA

Patricia Tueting
The Psychiatric Institute, Department of Psychiatry, College of Medicine, University of Illinois at Chicago, Chicago, Illinois, USA

Luis M. Valor
Instituto de Neurociencias, Universidad Miguel Hernández-Consejo Superior de Investigaciones Científicas, Alicante, Spain

PREFACE

This issue of *Progress in Molecular Biology and Translational Science* is focused on neuroepigenetics, a still relatively "young" area of research exploring the roles of chromatin structure and function in the context of development, adulthood, and disease. In the broadest sense, epigenetics could be viewed as the mechanisms and molecular bridges by which countless internal and external factors (re)organize the genomic material that is packaged inside the nuclei of brain cells, thereby potentially affecting synaptic plasticity and behavior. Epigenetic mechanisms, which among many others include proper regulation of DNA cytosine methylation, histone modifications, and nucleosomal organization (a nucleosome is the elementary unit of chromatin, with 147 bp of genomic DNA wrapped around a core of eight histone proteins), are thought to play an important role in various neurodevelopmental and neurological disorders.

Some of these key concepts linking epigenetic mechanisms to neuronal plasticity and (mal)adaptive mechanisms in the immature and also adult brain are highlighted in the eight chapters published in this issue. The first chapter, written by Jarome, Thomas, and Lubin, provides a general overview and detailed discussion on regulatory mechanisms governing covalent DNA and nucleosomal histone modifications during memory formation and storage in the mammalian brain. Next in Chapter 2, Rampp, Binder, and Provençal discuss gene × environment interactions in the context of traumatic memory and posttraumatic stress disorder, with particular focus on steroid hormone and other stress-responsive pathways. In Chapter 3, Hitchcock and Lattal describe the increasingly recognized importance of histone methylation and acetylation and other modifications for plasticity-dependent phenomena in the brain's reward circuit, and the potential implications for the treatment of addiction and substance abuse and dependence. Chapters 4, written by Guidotti, Dong, Tueting, and Grayson, and 5, written by Lolak, Suwannarat, and Lipsky, discuss epigenetic risk factors in the context of mood and psychosis spectrum disorders, including schizophrenia and depression. Next, Lopez-Atalaya, Valor, and Barco in Chapter 6 provide a highly informative discussion of Rubinstein–Taybi syndrome (RSTS), a neurodevelopmental syndrome caused by mutations in the genes encoding the lysine acetyltransferases CBP and p300. RSTS serves as a prototype example of monogenic neurological disease associated with epigenetic dysregulation (histone acetylation in this case), affecting both the developing brain and mature circuitry via multiple mechanisms. While aforementioned reviews focus on environmental or genetic mechanisms directly impacting the epigenome of brain cells, Chapter 7 by Galler and Galler Rabinowitz explores intergenerational effects of early adversity, including the potential role of epigenetics in intergenerational transmission. The final chapter of this issue, written by Mitchell, Roussos, Peter, Tsankova, and Akbarian, embarks on future developments in the field of neuroepigenetics (which will include the exploration of the nonrandomness in three-dimensional chromosomal organization and other principles often referred to as "higher order chromatin") and how such type of molecular approaches could provide further insights into genetic and epigenetic risk architecture of cognitive disorders unique to the human brain.

We, as volume editors, have asked the authors of these various chapters not only to summarize the evidence collected so far in their field but also to highlight some of the important topics that are subject to ongoing debate or hitherto underexplored. It is our hope and expectation that such type of discussion will further stimulate interest in epigenetic approaches and excite the younger generation of neuroscientists about these lines of research. Just to mention a few examples of debate mentioned in this issue: Chapter 1 highlights the bulk of the epigenetics literature pertaining to the learning and memory field and is preoccupied with mechanisms that are primarily relevant for initial acquisition of memory, while much less attention has been given to the equally important stages of consolidation, retrieval, and extinction. This chapter also highlights the potential epigenetic targets that may serve to improve or enhance memory function, while Chapter 5 cautions on the effects of neurological therapies on these epigenetic mechanisms. Several chapters include some remarks on tissue specificity of epigenetic signals, which is particularly relevant in the context of approaches aimed at exploring epigenetic signals in blood and the periphery as biomarkers for brain-related disease (Chapter 2) or chromatin state mappings in brain tissue with its extremely complex and highly heterogenous mixture of neuronal and glial subpopulations (Chapter 8). Furthermore, Chapter 3 points to two other important problems in neuroepigenetics including the often unresolved question of correlational association versus causal mechanism, and the difficulties and challenges in linking the molecular and behavioral phenotypes to the prevailing theories on reward and memory. Chapter 7 touches upon the still very novel and provocative concept of transgenerational inheritance.

We, as volume editors, are extremely grateful for the privilege to have worked with an outstanding group of colleagues. We would like to thank Ms. Helene Kabes for outstanding editorial assistance and the editors of the *PMBTS* series, Dr. Michael Conn and Ms. Mary Ann Zimmermann, for their support.

Schahram Akbarian
Farah Lubin

CHAPTER ONE

The Epigenetic Basis of Memory Formation and Storage

Timothy J. Jarome, Jasmyne S. Thomas, Farah D. Lubin
Department of Neurobiology, University of Alabama at Birmingham, Birmingham, AL, USA

Contents

Abstract

The formation of long-term memory requires a series of cellular and molecular changes that involve transcriptional regulation of gene expression. While these changes in gene transcription were initially thought to be largely regulated by the activation of transcription factors by intracellular signaling molecules, epigenetic mechanisms have emerged as an important regulator of transcriptional processes across multiple brain regions to form a memory circuit for a learned event or experience. Due to their self-perpetuating nature and ability to bidirectionally control gene expression, these epigenetic mechanisms have the potential to not only regulate initial memory formation but also modify and update memory over time. This chapter focuses on the established, but poorly understood, role for epigenetic mechanisms such as posttranslational modifications of histone proteins and DNA methylation at the different stages of memory storage. Additionally, this chapter emphasizes how these mechanisms interact to control the

Progress in Molecular Biology and Translational Science, Volume 128
ISSN 1877-1173
http://dx.doi.org/10.1016/B978-0-12-800977-2.00001-2

ideal epigenetic environment for memory formation and modification in neurons. The reader will gain insights into the limitations in our current understanding of epigenetic regulation of memory storage, especially in terms of their cell-type specificity and the lack of understanding in the interactions of various epigenetic modifiers to one another to impact gene expression changes during memory formation.

1. INTRODUCTION TO THE NEUROBIOLOGY OF LEARNING AND MEMORY

The process of encoding long-term memories in the brain for later recall is complex, requiring simultaneous engagement of neurons across multiple brain regions. The neuronal circuits recruited for this process can vary based on the type of behavioral task but include neurons in the hippocampus for processing of contextual and spatial information,[1] the amygdala for processing of emotional content,[2] and the insular cortex for processing of gustatory information.[3] For example, one of the most utilized behavioral paradigms for elucidating the molecular and cellular mechanisms of memory formation and storage is Pavlovian fear conditioning.[4] In this paradigm, a neutral conditioned stimulus (CS) is paired with a noxious unconditioned stimulus (UCS). Once paired, presentation of the CS by itself can elicit emotional and physiological responses similar to those produced by the aversive UCS. Due to the emotional component of the task, associations acquired using Pavlovian fear conditioning requires the amygdala for long-term memory formation and storage. However, some fear memories, such as those that are contextually based, also require the hippocampus, anterior cingulate cortex (ACC), and retrosplenial cortex, whereas other fear memories, such as those that are auditory-based, do not rely on these regions (reviewed in Ref. 5). Thus, memory storage is a dynamic process that requires synaptic plasticity in many different brain regions, though the memory-related brain regions required can vary greatly depending on the behavioral paradigm used.

Once acquired, memories can go through several different stages of modification and storage (Fig. 1.1). For example, following the acquisition of a fear conditioning task, there is a time-dependent process (<6 h) that occurs at the molecular level within the hippocampus and other brain regions, which is necessary to transfer labile short-term memories to stable long-term memories, a process referred to as memory consolidation.[6] Once

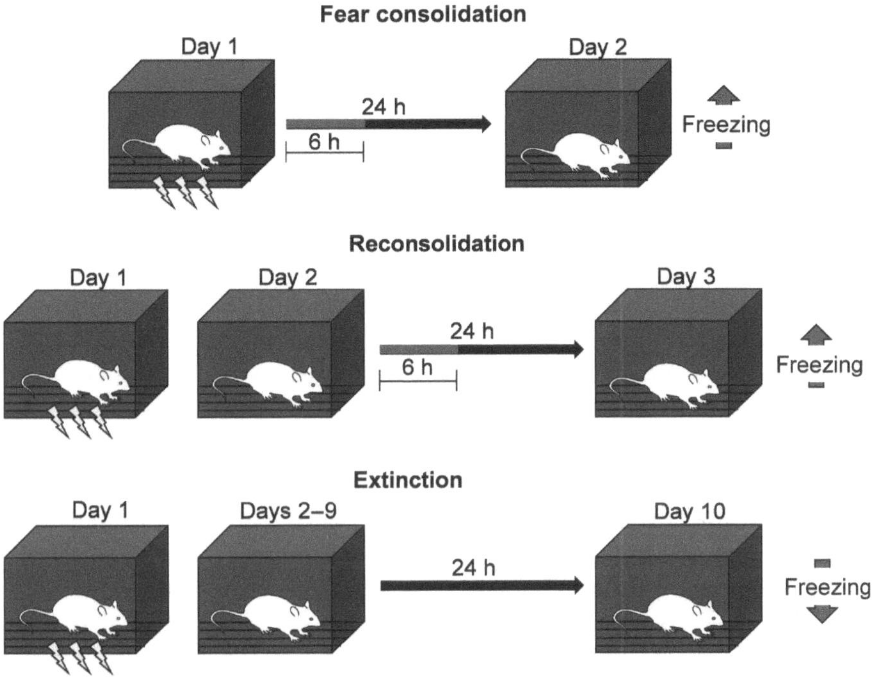

Figure 1.1 Memories go through several distinct phases of storage at the molecular level. In this example, a rat is trained to a contextual fear conditioning paradigm in which a novel context is paired with a mild aversive footshock. Following the training experience, the learned association (context plus footshock) will consolidate a process where gene transcription and protein synthesis occur. When the animal is reexposed to the training context in the absence of the footshock, it displays increased freezing behavior as an indication of memory retention. Additionally, upon retrieval (reexposure to the training context), the consolidated memory enters a new state of lability necessary for memory updating and modification, known as reconsolidation. This process maintains or increases fear on later tests. However, if the animal is exposed to the training context many times over without receiving the shock, the memory will extinguish, which results in a reduction in fear on later tests. Red lines denote theoretical time course of the consolidation and reconsolidation processes. (See the color plate.)

a memory is consolidated, it was generally believed to be no longer susceptible to disruption. However, upon retrieval, a once consolidated memory "destabilizes" and undergoes a new time-dependent molecular process (2–6 h) termed reconsolidation, which is necessary to return the memory to long-term storage.[7] This reconsolidation process is thought to be a unique phase in which memories for previously acquired associations or learned tasks can be modified.[8] Consistent with this, reconsolidation only occurs in the presence of new information and can be used to persistently attenuate memories for traumatic events,[9,10] suggesting that the reconsolidation process may have important clinical implications. Additionally, upon repeated exposure to the CS associated with the original learning experience without the presentation of the UCS, some memories can undergo extinction whereby the CS no longer elicits the emotional and physiological responses it acquired during the initial learning event. This "extinction" process is a form of new inhibitory learning that depends largely on the amygdala and infralimbic region of the medial prefrontal cortex and is often used in therapeutic settings to help reduce fear and emotional distress associated with memories for traumatic events.[11,12] Thus, memories for learned tasks and associations can go through several different stages of modification and storage that allow for it to be altered and persist over time.

For decades, the leading theory supporting the process of long-term memory formation and storage is that dynamic changes in gene transcription and protein synthesis in neurons are required.[13] Consistent with this, numerous studies have shown impairments in long-term memory for a variety of different behavioral tasks following transcriptional or translational blockade in neurons (reviewed in Ref. 5). Additionally, memory impairments are observed following pharmacological and genetic manipulations of signaling molecules thought to be upstream of transcription and translation, including NMDA receptor, protein kinase A, and calcium/calmodulin-dependent protein kinase II-dependent signaling (for review, see Johansen *et al.*[14]). Furthermore, behavioral training results in bidirectional changes in gene expression, suggesting that memory formation requires both increased and decreased gene transcriptions in neurons (e.g., Refs. 15,16). Collectively, these results support the theory that dynamic regulation of mRNA transcription is a critical step in memory formation and storage in neurons. However, while some transcription factors have been implicated in memory consolidation (reviewed in Ref. 17), very little is known about how up- or downregulation of gene transcription is altered in memory-related brain regions during long-term memory formation and storage. Recently, the field of "neuroepigenetics" has emerged as a viable mechanism to help further define how dynamic gene transcription changes are modified during the memory storage process. In this chapter, the reader will gain an understanding of the role of epigenetic mechanisms in transcriptional regulation of genes in neurons that are necessary for long-term memory formation and storage. Further, assuming that these epigenetic modifications are reversible, they can potentially serve as candidate targets for numerous memory disorders that arise from neurological conditions such as Alzheimer's disease, posttraumatic stress disorder, and epilepsy.

2. EPIGENETIC MECHANISMS OF MEMORY CONSOLIDATION

Traditional cellular models of memory consolidation theorized that gene transcription was regulated by intracellular signaling molecules following behavioral training.[18] However, epigenetic mechanisms such as acetylation and methylation of histone proteins and methylation of DNA have emerged as critical regulators of mRNA transcription during the memory consolidation process. Importantly, these epigenetic mechanisms have been shown to both enhance and repress gene expression in neurons, suggesting that epigenetic mechanisms can bidirectionally regulate gene transcription during memory formation. Considering that some of these epigenetic marks have been shown to persist beyond the memory consolidation period, it is possible that certain epigenetic mechanisms may act to not only consolidate memories but also maintain them over time in the face of continued mRNA and protein turnover. In this section, we review both known and potential epigenetic regulators of memory consolidation in neurons.

2.1. Histone acetylation

The most studied epigenetic mechanism in memory formation and storage is histone acetylation, which is thought to regulate an active transcriptional state necessary for memory consolidation in neurons. The first evidence that acetylation of histones was involved in memory formation came from a studying demonstrating that histone acetylation was increased in the insular cortex of mice following novel taste learning.[19] Consistent with this, numerous studies have reported increases in acetylation of different histones sites following learning (reviewed in Ref. 20). Histone acetylation opens the chromatin structure, exposing DNA and allowing increased transcription factor binding, resulting in increased gene expression.[21] Accordingly, several genes show increased histone acetylation levels of their promoter regions following learning and inhibiting histone acetylation alters expression of these genes. Thus, learning-induced increases in histone acetylation provide a powerful mechanism by which gene transcription can be increased in neurons.

The ability of histones to become acetylated is controlled by histone lysine acetyltransferases (H/KATs) and histone lysine deacetylases (H/KDACs), which work in concert to promote the ideal epigenetic setting for memory consolidation. It would be expected then that memory consolidation would require a balance between activations of H/KATs and H/KDACs to appropriately increase histone acetylation in neurons. Indeed, numerous studies have found that preventing increases in histone acetylation by pharmacologically or genetically inhibiting histone acetyltransferase (HAT) activity blocks memory consolidation, resulting in long-term memory impairments for a variety of behavioral tasks (reviewed in Ref. 22). Additionally, many studies have found that manipulation of H/KDACs can have profound effects on synaptic plasticity and memory consolidation. For example, genetic deletion of H/KDAC2 in neurons enhances memory consolidation, while overexpression of H/KDAC2 decreases dendritic spine density, synapse number, and synaptic plasticity and impairs memory formation.[23] Results such as these suggest that not only is histone acetylation a critical epigenetic regulator of memory formation but also H/KDACs act to limit memory strength by controlling the amount of histone acetylation present in neurons following learning.[24] Consistent with this, enhancing histone acetylation with H/KDAC inhibitors can result in the formation of an object location memory that persists well beyond what is typically normal for that behavioral paradigm.[25] Thus, proper memory consolidation requires very precise, highly regulated changes in histone acetylation, which increase gene transcription necessary for memory formation and storage, and H/KDACs act as a way to limit the overall amount of histone acetylation, constraining memory strength. It is important to note, however, that H/KDACs can also target other nonhistone proteins, a mechanism unexplored in the context of memory formation and storage.[26] Thus, H/KDACs may regulate memory strength through regulation of acetylation of both histone and nonhistone proteins.

2.2. Histone methylation

In addition to acetylation, histone methylation has also been implicated in the memory consolidation process, though transcriptional regulation mediated by this epigenetic mechanism is much more complex (Fig. 1.2). This is due to the diverse methyl modifications that histones can undergo as lysine residues can be mono-, di-, or trimethylated while arginine residues can be mono- or dimethylated and the type of modification the site undergoes can

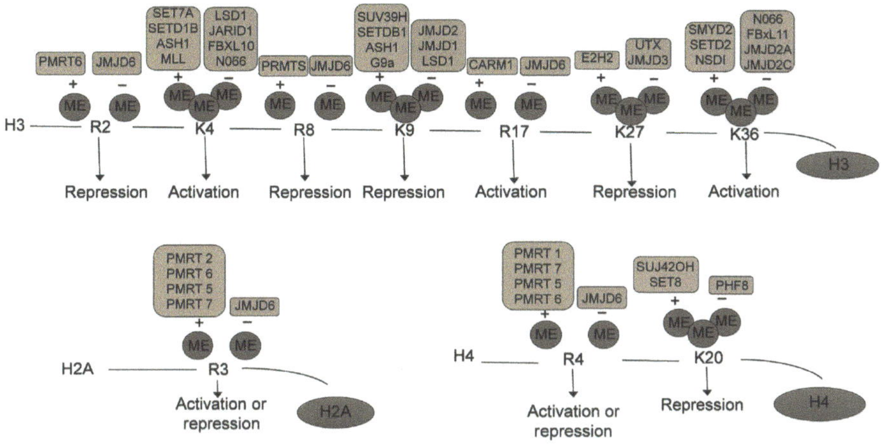

Figure 1.2 Coordinated regulation of histone methylation dynamically regulates different transcriptional states. Histones can be methylated at different arginine (R) and lysine (K) sites, which are regulated by specific sets of histone methyltransferases (green boxes) and demethylases (blue). Arginine residues can be mono- or dimethylated, while lysine residues can be mono-, di-, or trimethylated, and this results in differential activation or repression of gene transcription. (See the color plate.)

have differential effects on gene transcription and the memory consolidation process.[27] For example, contextual fear conditioning globally increases histone H3 lysine 9 dimethylation (H3K9me2), a transcriptionally repressive mark, and histone H3 lysine 4 trimethylation (H3K4me3), a transcriptional activator, in the hippocampus and entorhinal cortex of rats.[15,28] Interestingly, these marks continue to be temporally regulated 24 h after behavioral training, suggesting that histone lysine methylation may continue to regulate gene transcription even after the consolidation process has completed. This is particularly important because so far no other epigenetic mechanism has been shown to persist beyond the memory consolidation period. Currently, the functional significance of these persistent changes in histone methylation remains unknown, but it is possible that they are important for continued control over the transcription of specific genes to allow stored memories to remain stable over time in the face of continued protein and mRNA turnover.[27]

Histone methylation is tightly regulated by specific histone lysine methyltransferases (H/KMTs) and histone lysine demethylases (H/KDMs), which promote or inhibit histone methylation, respectively. The histone methyltransferases can be very selective, targeting a specific amino acid residue, modifying that site with a precise number of methyl modifications that result in different transcriptional states. For example, the H/KMT G9a controls H3K9me2 while the H/KMT-MLL regulates H3K4me3. Interestingly, genetic deletion of H/KMT-MLL or pharmacological blockade of H/KMT-G9a in the hippocampus impairs fear memory consolidation.[15,16,28] Additionally, pharmacological manipulation of the H/LDM-LSD1 in the hippocampus can enhance fear memory, suggesting that HDM may act to limit memory strength. This suggests that increases and decreases in gene transcription mediated by histone lysine methylation are critical for long-term memory formation in the hippocampus. Consistent with this, microarray analyses have revealed that learning in a fear conditioning task results in dynamic, bidirectional changes in gene expression in the hippocampus.[15] Some of these changes in gene expression are associated with increased H3K4me3 binding of their promoter regions that correlates with increased DNA methylation, suggesting a relationship between histone methylation and DNA methylation. Additionally, the repression of some genes following behavioral training is associated with increased H3K9me2 binding of their promoter regions, and learning-induced changes in H3K9me2 can be reversed by an H/KDAC inhibitor,[28] suggesting that histone acetylation and methylation may interact to regulate transcription during long-term memory formation. Collectively, this indicates that histone methylation regulates changes in gene transcription in neurons necessary for memory consolidation. In fact, genetically inhibiting the H/KMT-MLL results in a dramatic loss of learning-dependent changes in gene expression in the hippocampus, suggesting that a single histone methylation modification can regulate the transcription of hundreds of different genes during memory consolidation.[16] However, inhibition of the H/KMT-G9a in the entorhinal cortex enhances long-term memory, while inhibition of the H/KDM-LSD1 impairs memory, suggesting that in some cases histone methylation may act to limit memory strength. Regardless, results such as these support the theory that histone methylation acts as an important regulator of gene transcription during the consolidation process.

2.3. Histone phosphorylation

In addition to modification by acetyl and methyl groups, histones can also be phosphorylated. This type of posttranslational modification is associated with an active transcriptional state, and, similar to histone acetylation and methylation, histone phosphorylation has been implicated in memory consolidation. For example, H3 phosphorylation increased in the CA1 region of the hippocampus following contextual fear conditioning.[29] This increase was dependent on NMDA receptor and ERK/MAPK (extracellular signal-regulated kinase/mitogen-activated protein kinase) signaling, two important regulators of the consolidation process,[30,31] suggesting that histone phosphorylation is increased in an activity-dependent manner. Consistent with this, mice lacking MSK1 (mitogen- and stress-activated protein kinase 1) show deficits in learning-induced increases in histone phosphorylation in the hippocampus.[32] These results provide convincing evidence that increases in histone phosphorylation correlate with altered behavioral performance; however, it is currently unknown if this epigenetic mechanism is actually a critical regulator of gene transcription during the consolidation process. It is conceivable that the observed increases in histone phosphorylation following behavioral training are a by-product of alterations in intracellular signaling and gene transcription. In this scenario, directly manipulating histone phosphorylation would not have an effect on the normal consolidation process, and memory would persist on later tests. Consequently, more research is needed to understand the exact role of histone phosphorylation in the regulation of gene transcription during the consolidation process.

2.4. Histone ubiquitination and sumoylation

While very little is known about the role of histone phosphorylation in the memory consolidation process, even less is known about the role of the protein modifiers ubiquitin and SUMO (small ubiquitin-like modifier) in transcriptional regulation mediated by histone modifications. The regulation of transcription by histone ubiquitination is complex, as it can both increase and decrease gene transcription depending on the site of modification. For example, the two most common forms of histone ubiquitination are histone H2B lysine 120 monoubiquitination (H2BubiK120) and histone H2A lysine 119 monoubiquitination (H2AubiK119), which promote and inhibit transcription, respectively. The ubiquitination of histones is regulated by protein ligases involved in the ubiquitin–proteasome system. Ubiquitin E2 and E3 ligases are involved in the conjugation of ubiquitin modifiers to histones, while deubiquitinating enzymes remove these marks with high affinity. Thus, histone ubiquitination is tightly regulated in a similar fashion as histone acetylation and methylation. While numerous studies have implicated degradation-specific ubiquitin signaling in the memory consolidation process (e.g., Ref. 33), it is currently unknown if learning even alters histone ubiquitination levels in neurons. In fact, to date, the importance of histone ubiquitination in transcriptional regulation has remained largely unexplored in both the contexts of memory storage and synaptic plasticity. This is particularly surprising considering some histone ubiquitination marks are thought to regulate histone "cross talk" in cell cultures *in vitro*[34] and histone marks are thought to be interconnected during the memory consolidation process (e.g., Ref. 28). This would suggest that histone ubiquitination could serve as a critical regulator of histone modifications at specific gene promoters, properly regulating gene transcription during long-term memory formation; however, this has never been examined. Considering the dynamic nature of ubiquitin signaling in long-term memory formation, the tight regulation of the ubiquitin process, and evidence that it can regulate histone "cross talk," the importance of histone ubiquitination in memory consolidation proves to be one of the most potentially promising areas of future epigenetic research examining transcriptional regulation during memory formation and storage.

Histone sumoylation, which is associated with an active transcriptional state, is similar to ubiquitination in that very little is known about the role of this epigenetic mechanism in transcriptional regulation during memory consolidation and synaptic plasticity. The protein modifier SUMO is attached to histones by ubiquitin ligases in the same manner that histones become ubiquitinated; however, unlike ubiquitination, protein sumoylation does not always require the presence of an E3 ligase. This histone modification is reversible as SUMO-specific proteases can remove the SUMO modification, suggesting that histone sumoylation-mediated changes in gene transcription are tightly regulated by ubiquitin protein ligases and SUMO proteases. To date, there is very little evidence implicating a role of protein sumoylation in long-term memory formation; however, one study did report changes in histone sumoylation at specific gene promoters in response to learning,[35] suggesting that this epigenetic mechanism may potentially contribute to transcriptional regulation during memory consolidation.

2.5. DNA methylation

While there is convincing evidence that posttranslational modification of histones contributes to transcriptional regulation during the consolidation process, a large number of studies have also suggested an important role for DNA methylation in memory storage (reviewed in Ref. 20). DNA methylation occurs at CpG sites when DNA methyltransferases (DNMTs) transfer methyl groups from the donor *S*-adenosylmethionine to the 5′ position of the cytosine pyramidal ring. Methylated CpG sites are associated with transcriptional silencing, and this methylation can be self-perpetuating, suggesting that DNA methylation could serve as a powerful mechanism of information storage in neurons across the life span. In fact, this idea has led to the theory that DNA methylation may serve as a mechanism of maintaining memories in the brain.[36] Consistent with this, numerous studies have found impairments in long-term memory for a variety of behavioral tasks following genetic or pharmacological manipulations of DNMTs.[37,38] This result suggests that DNA methylation may be a critical regulator of gene expression during long-term memory formation. Indeed, DNA methylation has been shown to dynamically regulate gene transcription during the memory consolidation process. For example, learning in a contextual fear conditioning paradigm results in dynamic changes in *Bdnf* (brain-derived neurotrophic factor) DNA methylation.[39] The *Bdnf* gene structure has multiple initiation sites that allow for isoform-specific transcription of mRNA transcripts. Interestingly, fear conditioning resulted in decreases in DNA methylation levels associated with *Bdnf* exons I and IV and increases at exon VI in the CA1 region of the hippocampus. These changes in DNA methylation were associated with exon-specific changes in *Bdnf* gene expression, and pharmacologically inhibiting DNMT activity prevented these changes in gene expression and impaired long-term memory, suggesting that regulation of the *Bdnf* gene by DNA methylation is potentially a critical mechanism of the memory consolidation process. Additionally, some recent evidence suggests that active DNA demethylation may be an important regulator of long-term memory formation as knockdown of Tet1 (ten-eleven translocation 1) expression alters memory for a fear conditioning task.[40] Results such as these suggest that DNA methylation can dynamically regulate gene transcription and expression in neurons necessary for the memory consolidation process.

As mentioned above, a currently popular theory is that DNA methylation could serve as a cellular mechanism of maintaining memory across the life span. However, while many studies have reported increases in DNA methylation following learning and that manipulation of DNMT activity alters long-term memory, to date, the scientific community has largely focused on changes in DNA methylation *during* the memory consolidation period. If DNA methylation does act as a mechanism of memory "maintenance," then you would expect to observe long-term changes in

DNA methylation that last well beyond the memory consolidation time period and that manipulation of this epigenetic mechanism after completion of the consolidation process would impair long-term memory. However, few studies to date have reported the effect of delayed DNMT inhibition on long-term memory. Currently, the only evidence implicating DNA methylation in memory maintenance is a study that found that contextual fear conditioning persistently increased DNA methylation of the *Calcineurin* gene in the ACC region of the prefrontal cortex.[41] Importantly, infusions of DNMT inhibitors in the ACC 30 days after fear conditioning impaired long-term memory, suggesting that DNA methylation maintained this type of memory in the prefrontal cortex. Surprisingly, despite increases in DNA methylation in the prefrontal cortex 1 day after contextual fear conditioning, infusions of DNMT inhibitors into the ACC 1 day after training did not impair long-term memory. This suggests that while DNA methylation may maintain memory under certain circumstances, it does not seem that this occurs shortly after the consolidation process has ended. Evidence such as this makes understanding the potential role of DNA methylation in memory maintenance after consolidation difficult, as a maintenance mechanism would be expected to be involved in memory storage persistently following the completion of the consolidation process; however, few molecules meet this requirement. In fact, while still controversial, the only molecule shown to be persistently involved in memory maintenance following completion of the consolidation process is the atypical (protein kinase C) isoform PKMζ (e.g., Refs. 42–45). Consequently, while the theory that DNA methylation may regulate both the consolidation and maintenance of long-term memories is intriguing, there is very little direct evidence, to date, to support this theory.

2.6. Summary of epigenetic regulation of memory consolidation

The studies reviewed in this section provide convincing evidence that epigenetic mechanisms underlie the process of memory consolidation (Fig. 1.3). Histone modifications such as methylation bidirectionally regulate changes in gene expression during memory formation, while histone acetylation is associated with transcriptional activation of gene expression and DNA methylation is largely associated with transcriptional repression.[46,47] In regard to DNA methylation, there are some cases wherein DNA methylation has been associated with increased gene transcription in neurons, as is the case of *zif268* in the CA1 region of the hippocampus

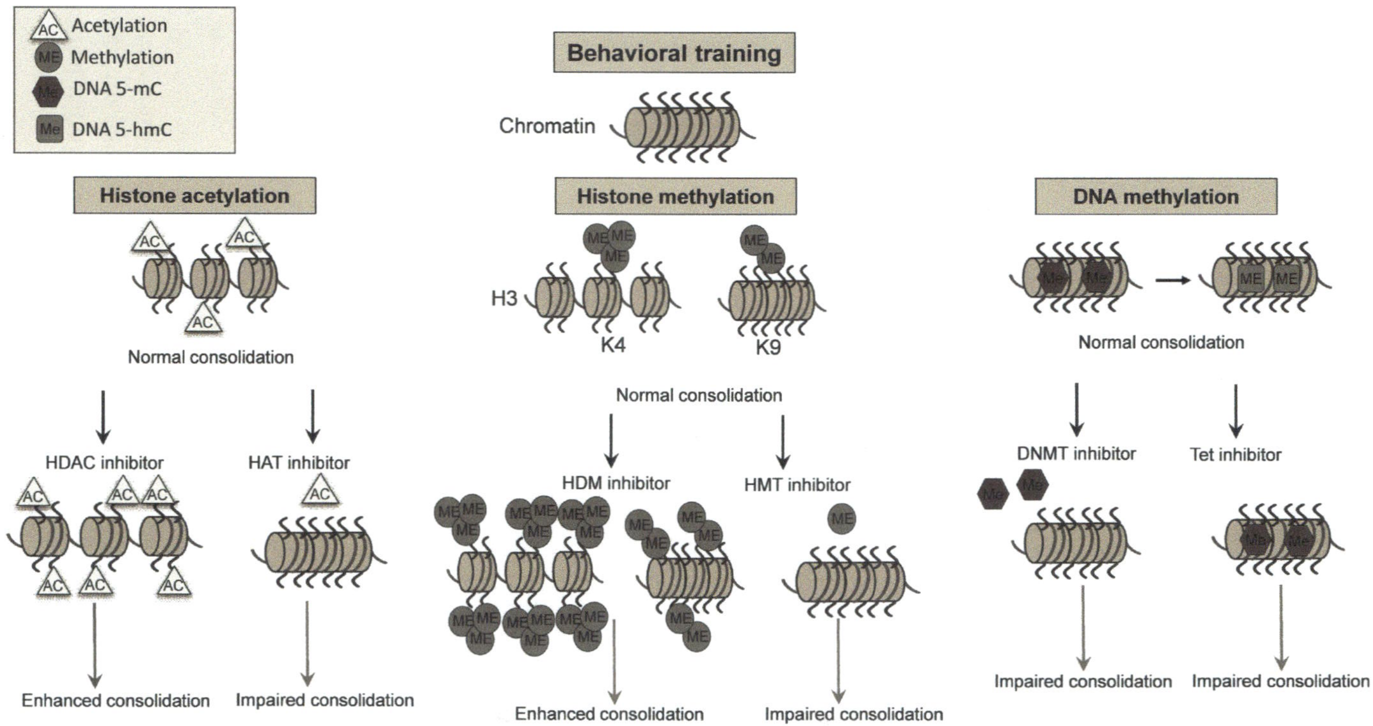

Figure 1.3 Epigenetic regulation of memory consolidation. Histone acetylation and histone methylation are increased in neurons following behavioral training. Inhibiting histone acetyltransferases (HATs) and histone methyltransferases (HMTs) impairs memory consolidation, while inhibiting histone lysine deacetylases (H/KDACs) and histone demethylases (HDMs) enhances memory. DNA methylation is increased following behavioral training in the form of 5-methylcytosine (5-mC), which is converted to 5-hydroxymethylcytosine (5-hmC), potentially promoting DNA demethylation. Inhibiting DNA methyltransferases (conjugate 5-mC) or Tet family proteins (convert 5-mC to 5-hmC) impairs memory consolidation. (See the color plate.)

following contextual fear conditioning.[28] Additionally, histone and DNA methylation may be involved in memory maintenance following the completion of the consolidation process; however, very few studies have examined this to date. It is important to note that these histone and DNA modifications are likely not isolated events but rather influence each other, as is the case of histone acetylation and methylation.[28] Future studies are necessary to better understand these coordinated changes in posttranslational modification of histones and DNA methylation that serve as a critical regulator of memory consolidation.

3. EPIGENETIC MECHANISMS OF MEMORY RECONSOLIDATION

While memories were once thought to be stable and no longer susceptible to disruption following the completion of the consolidation process, in recent years, there has been a renewed interest in the theory that memories can again become labile following retrieval, a process referred to as reconsolidation.[48] This reconsolidation process is a unique stage in which memory content can be modified and requires *de novo* gene transcription and protein translation in neurons (reviewed in Ref. 5). Similar to the consolidation process, evidence suggests that epigenetic mechanisms may regulate this need for new mRNA transcription following retrieval; however, in general, much less is known about the mechanisms of transcriptional regulation during memory reconsolidation. In this section, we review both known and potential epigenetic mechanisms of the reconsolidation process.

3.1. Histone modifications during memory reconsolidation

The first study to implicate epigenetic mechanisms in transcriptional regulation during the reconsolidation process found that retrieval of a contextual fear memory increased global H3 acetylation in the CA1 region of the hippocampus.[49] These global changes were associated with gene-specific increases of histone acetylation of the *zif268* and *IκBα* promoters, suggesting that histone acetylation may regulate increases in gene expression during the reconsolidation process, and were dependent on the NF-κB (nuclear factor-kappa B) signaling pathway. Importantly, inhibiting H/KDAC activity did not alter long-term memory but did rescue memory impairments that normally resulted from NF-κB inhibition in neurons following memory retrieval, suggesting that H/KDACs act to regulate histone acetylation and memory storage during the reconsolidation process. Consistent with this, H/KDAC inhibitors can rescue memory impairments that result from blocking NF-κB signaling in the amygdala following memory retrieval.[50] These results suggest that histone acetylation may be an important regulator of gene transcription during the reconsolidation process. Accordingly, inhibiting HAT activity impairs the reconsolidation of fear memories in the amygdala, while pharmacological manipulation of H/KDAC activity can result in memory strengthening following retrieval.[51,52] Collectively, these results suggest that coordinated regulation of histone acetylation is a critical regulator of memory reconsolidation following retrieval. However, while histone acetylation may be critical for the reconsolidation process, very little is known about how it may regulate gene transcription following memory retrieval. For example, the *zif268* and *IκBα* gene promoters are the only ones known to undergo changes in histone acetylation levels, and it is unknown if manipulation of histone acetylation alters changes in gene expression during the reconsolidation process. Thus, while it is clear that histone acetylation is a critical regulator of the reconsolidation process, very little is known about how it regulates *de novo* gene transcription following retrieval.

While evidence clearly demonstrates a role for histone acetylation in the regulation of the reconsolidation process, very little is known about the role of other histone modifications in this stage of memory storage and modification. For example, histone H3 phosphorylation is increased in the amygdala and the CA1 region of the hippocampus following the retrieval of a contextual fear memory; however, it is unknown if this epigenetic mechanism is critical for the reconsolidation process. Similar to memory consolidation then, the role of histone phosphorylation in transcriptional regulation during memory reconsolidation remains equivocal. Additionally, it is unknown if histone methylation is altered during the reconsolidation process, which is surprisingly considering that persistent changes in histone methylation have been observed in the hippocampus and entorhinal cortex following learning in a contextual fear conditioning paradigm.[15] Considering that reconsolidation is thought to occur in the same neurons in which the memory consolidated,[53] it would be expected that histone methylation may be a critical bidirectional regulator of gene transcription during the reconsolidation process, though, to date, this has not been examined. It is also currently unknown if histone ubiquitination is involved in memory reconsolidation. While this may be expected considering the role of histone ubiquitination in memory consolidation is unknown, it is actually surprising because there is a well-documented role for degradation-specific ubiquitin signaling in reconsolidation in neurons.[33,54] In fact, ubiquitin signaling is theorized to be a master regulator of the reconsolidation process.[5,55] As a result, we currently know very little about how histone modifications epigenetically regulate gene transcription during the memory reconsolidation process.

3.2. DNA methylation during memory reconsolidation

While there is strong support for DNA methylation in transcriptional regulation during memory consolidation, very few studies have examined the same requirement for DNA methylation in the reconsolidation process. One study found that infusions of a DNMT inhibitor into the amygdala impaired the reconsolidation of an auditory fear memory.[56] Additionally, DNMT inhibitors can reverse learning-induced changes in auditory-evoked field potentials in the amygdala when infused around the time of memory retrieval, suggesting that DNA methylation is an important regulator of retrieval-induced memory lability at the cellular level.[57] However, while DNA methylation has been shown to be an important regulator of the reconsolidation process, it is unknown how DNA methylation regulates transcriptional processes and, more importantly, what genes are regulated by this epigenetic mechanism following retrieval. Additionally, it is unknown if DNA methylation is required in brain regions other than the amygdala during the reconsolidation process. As a result, we currently have only a basic understanding of the role of DNA methylation in the reconsolidation process.

3.3. Epigenetic regulation of reconsolidation-dependent memory updating

One of the most unique features of the reconsolidation process is that it allows a previously consolidated memory to be "updated" or have the precise content of the memory changed. One of the best known examples of this is the reconsolidation-update procedure, in which an isolated retrieval event is followed by a "massed" session of numerous retrieval events (e.g., 20) 10 min to 2 h later. This unique spacing of what is essentially a single retrieval trial followed by a large group of retrieval trials results in the persistent attenuation of fear responses to cues associated with traumatic events.[10] What is most intriguing about this paradigm is that it is effective in both rats and humans, suggesting that the reconsolidation-update procedure could have major clinical implications for the treatment of fear and anxiety associated with memory for traumatic events.[58] While very little is known about the cellular mechanisms that regulate this "updating" or "erasure" of fear memories, histone acetylation has also been shown to be involved in the reconsolidation-dependent updating of retrieved fear memories using the reconsolidation-update procedure.[59] This study found that retrieval of a contextual fear memory increased histone acetylation in the CA1 region of the hippocampus; however, this only occurred when the memory was retrieved 1 day but not 30 days after training. This is consistent with evidence suggesting that the hippocampus plays a time-limited role in the storage of contextual fear memories.[60] Interestingly, reconsolidation-update procedures are only effective at attenuating fear to recent, but not remote, memories,[61] suggesting that the effectiveness of the reconsolidation-update procedure correlates with the "systems-consolidation" process. Remarkably, infusions of a H/KDAC inhibitor into the CA1 region facilitated the reconsolidation-dependent updating of a retrieved remote fear memory (i.e., 30 days old), suggesting that H/KDACs may act to limit the updating of older fear memories. Collectively, these results suggest that epigenetic mechanisms regulate not only normal reconsolidation but also the reconsolidation-dependent updating of fear memories. However, to date, this is the only study that has examined the role of epigenetic mechanisms in the regulation of gene transcription during memory updating. Hence, much work is required to completely understand how epigenetic mechanisms contribute to memory updating following retrieval.

3.4. Summary of epigenetic regulation of memory reconsolidation

While there is strong evidence that histone modifications and DNA methylation regulate gene transcription during the consolidation process, very few studies have implicated epigenetic mechanisms in the regulation of memory reconsolidation. In Fig. 1.4, a summary of the known epigenetic mechanisms involved in memory reconsolidation is shown. Additionally, we have limited understanding of how these epigenetic mechanisms serve to regulate gene transcription following memory recall. As a result, we are only in the early stages of understanding how reconsolidation is epigenetically regulated. Despite this, it is clear that epigenetic regulation of gene expression is likely an important step in the reconsolidation process and may be a mechanism that can prove beneficial for reconsolidation-based

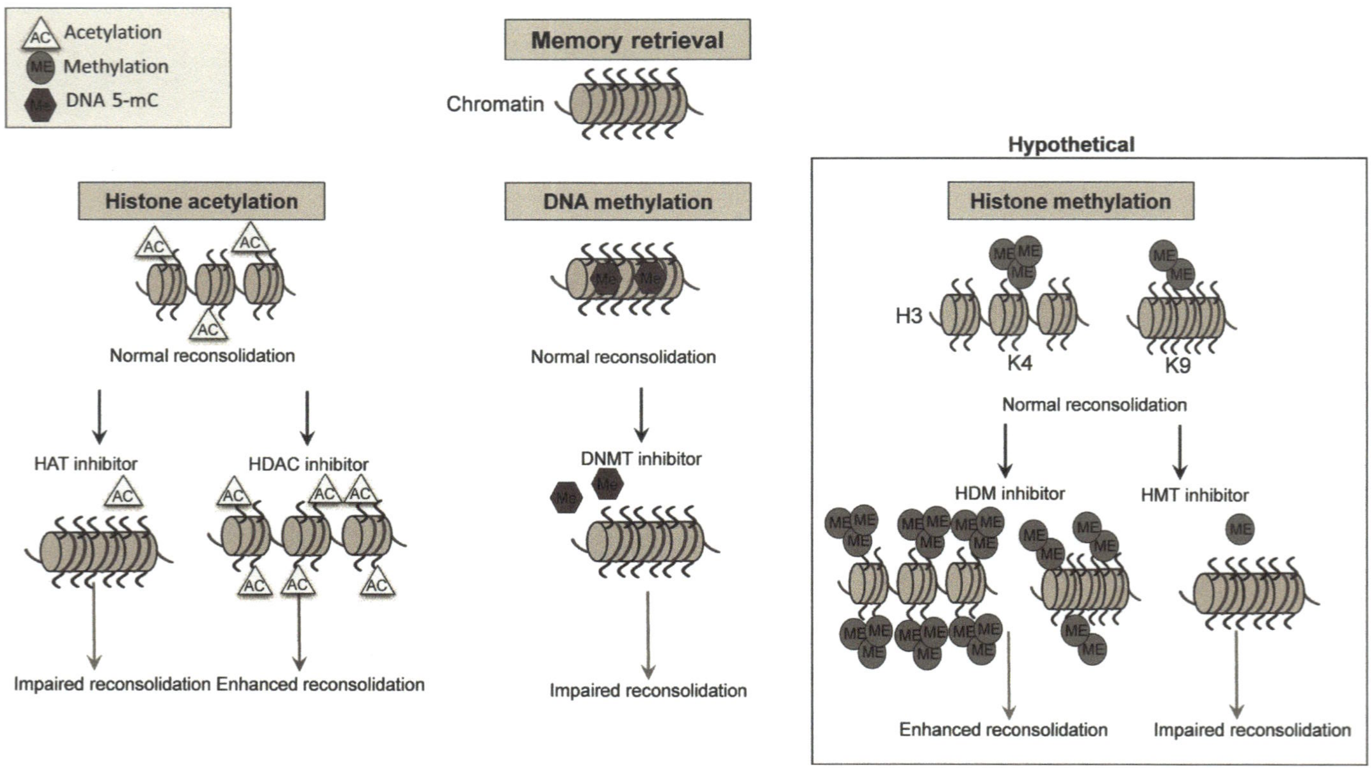

Figure 1.4 Epigenetic regulation of memory reconsolidation. Histone acetylation and DNA methylation (5-methylcytosine, 5-mC) are increased in neurons following memory retrieval. Inhibiting histone acetyltransferases (HATs) and DNA methyltransferases (DNMTs) impairs memory reconsolidation, while inhibiting histone lysine deacetylases (H/KDACs) enhances memory. Histone methylation is a hypothetical mechanism of the reconsolidation process, where memory retrieval increases histone methylation and inhibiting histone methyltransferases (HMTs) impairs reconsolidation while inhibiting histone demethylases (HDMs) enhances memory following retrieval. (See the color plate.)

therapeutic strategies. Therefore, future studies will likely look to better elucidate the epigenetic mechanisms of the reconsolidation process.

4. EPIGENETIC MECHANISMS OF MEMORY EXTINCTION

While a single retrieval trial initiates the memory reconsolidation process, repeated retrieval trials that occur consecutively or over a number of days result in memory extinction. This extinction process was once thought of as a way of erasing memories, but it is now believed to be due to the formation of a new inhibitory memory that supersedes the previously acquired excitatory memory.[11] However, this process results in a reduction in fear and anxiety with memories for traumatic events, suggesting that the extinction process can have therapeutic utility. Like consolidation and reconsolidation, extinction consolidation requires NMDA receptor-mediated plasticity including gene transcription and protein translation changes in neurons. However, while numerous studies have done a nice job identifying the brain regions involved in the extinction of memories, very little is known about how gene transcription is regulated during the extinction consolidation process.

Recently, epigenetic mechanisms have emerged as potentially important regulators of extinction consolidation in neurons. For example, extinction of an auditory fear memory increased histone H4 acetylation of the *Bdnf* exon IV promoter in the prefrontal cortex, and injection of the H/KDAC inhibitor valproic acid enhanced memory extinction, suggesting that histone acetylation was an important regulator of gene transcription during the extinction consolidation process. Interestingly, the H/KDAC inhibitor also enhanced H4 acetylation of *Bdnf* exon IV and gene expression when given prior to a weak extinction protocol, suggesting that H/KDACs act to limit the extinction process.[62] Consistent with this, H/KDAC inhibitors have been shown to enhance the extinction of both fear and drug-associated memories (reviewed in Ref. 63). Since H/KDACs limit memory extinction, then it would be expected that HATs are required for the extinction consolidation process. Surprisingly though, inhibiting activity of the HAT p300/CBP in the infralimbic region of the prefrontal cortex actually enhances extinction consolidation and can even do so for weak extinction protocols.[64] Despite this, enhancing activity of the HAT PCAF (p300/CBP-associated factor) in the infralimbic cortex enhances extinction consolidation,[65] suggesting that some HATs can limit memory extinction while others promote it. Regardless, these results provide strong evidence that histone acetylation is an important regulator of gene transcription during the extinction consolidation process.

Similar to memory reconsolidation, it is currently unknown if other histone modifications such as methylation, phosphorylation, and ubiquitination are involved in the extinction process in neurons. Additionally, it is unknown if DNA methylation contributes to transcriptional regulation during this phase of memory modification and storage. However, recent evidence suggests that Tet1 is critical for the extinction of fear memories.[66] Since Tet1 is thought to promote DNA demethylation through the conversion of 5-methylcytosine (5-mC) to 5-hydroxymethylcytosine (5-hmC), it is possible then that DNA methylation and demethylation are important regulators of the extinction consolidation process, though this has never been directly tested. Nonetheless, these results suggest that epigenetic mechanisms regulate gene transcription during memory extinction. In Fig. 1.5, we summarize the epigenetic mechanisms shown to be involved in memory extinction including histone acetylation and DNA methylation. Because histone methylation is closely associated with histone acetylation and DNA methylation regulation, a hypothetical schematic representation of the role that histone methylation plays in memory extinction is also presented.

5. FUTURE DIRECTIONS AND CONCLUSIONS

The studies described here provide strong evidence that epigenetic mechanisms regulate gene transcription during long-term memory formation in storage. Importantly, these mechanisms result in bidirectional changes in gene expression, which suggest that proper memory storage requires repression as much as it requires enhancements in gene transcription. However, while the field of neuroepigenetics has advanced rapidly in the last decade, there is still much that is not known about the epigenetic regulation of memory storage. First, several epigenetic modifications, such as histone ubiquitination and sumoylation, remain largely unexplored in the context of memory formation and synaptic plasticity. While these two modifications may not be as robust as histone acetylation and methylation, they provide an important means of bidirectional transcriptional regulation during activity-dependent synaptic plasticity such as memory formation and could provide important information about the epigenetic regulation of a subset of genes. Second, very little is known about how different epigenetic mechanisms interact with each other to regulate the transcription of specific

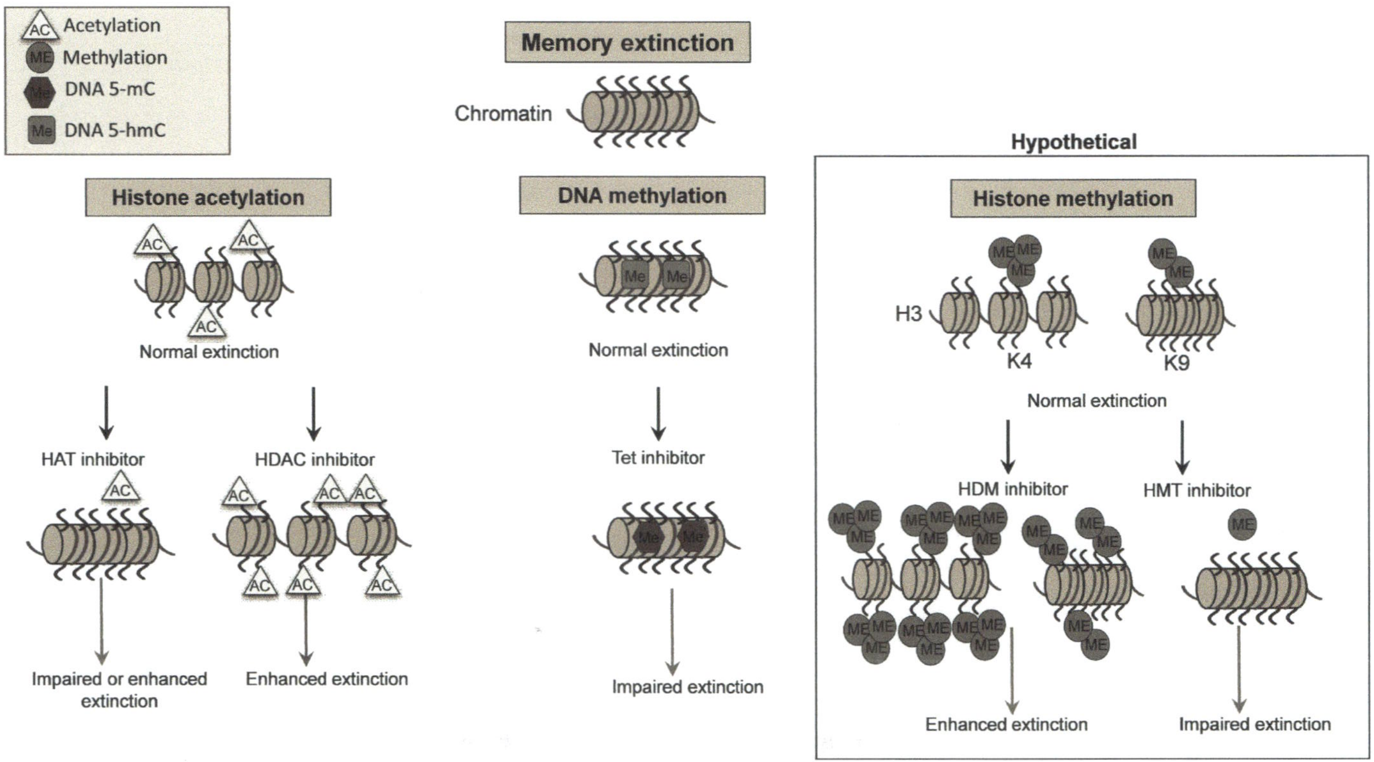

Figure 1.5 Epigenetic regulation of memory extinction. Histone acetylation and DNA methylation (5-hydroxymethylcytosine, 5-hmC) are increased in neurons following extinction training. Inhibiting histone acetyltransferases (HATs) can enhance or impair memory extinction, while inhibiting histone lysine deacetylases (H/KDACs) enhances extinction. Inhibiting the Tet family of proteins (convert 5-mC to 5-hmC) impairs memory extinction. Histone methylation is a hypothetical mechanism of the extinction process, where extinction training increases histone methylation and inhibiting histone methyltransferases (HMTs) impairs extinction consolidation while inhibiting histone demethylases (HDMs) enhances memory extinction. (See the color plate.)

genes necessary for memory storage. Some evidence has indicated that histone acetylation, methylation, and DNA methylation may work in concert to appropriately regulate gene transcription during memory formation[15,28,67]; however, few studies have directly examined the potential relationship between these modifications in the regulation of transcription for a given gene. Because histone methylation is intricately linked to the regulation of histone acetylation and DNA methylation, in Fig. 1.6, we illustrate these relationships that may be involved in transcriptional regulation of genes during memory formation. Additionally, very little is known about the cell type-specific epigenetic mechanisms that are involved in memory formation. For example, pyramidal neurons in the CA1 region of the hippocampus are thought to be preferentially involved in memory formation and storage. However, it is currently not known if epigenetic changes occur specifically in these cells following learning or if other cell types, such as astrocytes, are the primary cells expressing these modifications. Thus, while the neuroepigenetics field is rapidly advancing and our knowledge of the

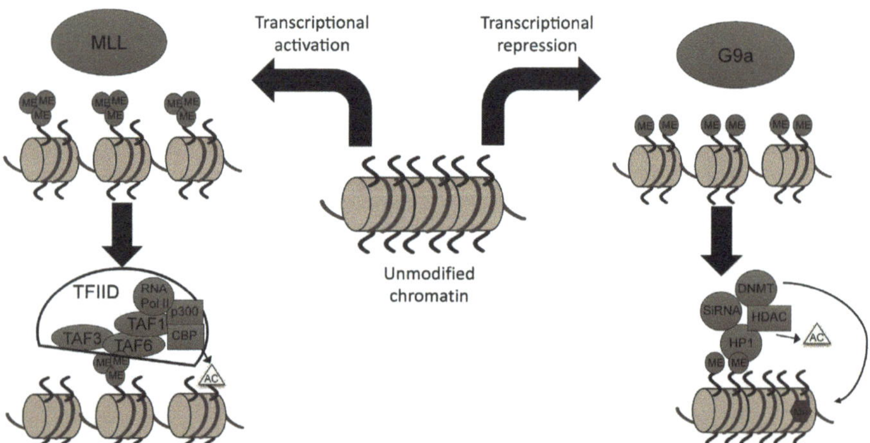

Figure 1.6 Histone methylation mediated homeostasis regulation of gene transcription during memory formation. Chromatin can be modified by the histone methyltransferase (HMT)-MLL and G9a to activate or repress gene transcription through the regulation of histone acetylation and DNA methylation. HMT-MLL trimethylates histone H3 lysine 4, recruiting a transcriptional complex containing RNA polymerase II and the histone acetyltransferases p300 and CBP, which increases histone acetylation, opens the chromatin structure, and enhances gene transcription. HMT-G9a dimethylates histone H3 lysine 9, recruiting a transcriptional complex containing histone lysine deacetylases (H/KDACs) and DNA methyltransferases (DNMTs), which closes the chromatin structure and results in transcriptional silencing. (See the color plate.)

epigenetic regulation of gene transcription during memory formation and storage has increased dramatically within the past decade, there are many questions remaining to be answered with regard to understanding the role of epigenetic mechanisms in memory consolidation, reconsolidation, and extinction.

One of the most intriguing findings from the studies reviewed here is that there exist close interactions between specific epigenetic modifications that function to limit memory strength. This is evident for both histone acetylation and methylation, as manipulation of the enzymes that remove or prevent these modifications (H/KDACs and HDMs, respectively) often enhances long-term memory retention. This lends the intriguing hypothesis that relieving this "molecular brake pad" could result in enhanced recollection and even promote memory formation for poorly learned tasks or associations.[24] Such an idea could have significant implications for promoting not only memory retention across time but also the permanent erasure of fear memories.[59] Future studies will unquestionably further examine this likely cognitive enhancement strategy to better determine its therapeutic utility.

The reconsolidation and extinction phases of memory modification and storage are widely theorized as a way of modifying (reconsolidation) and eliminating (reconsolidation and extinction) memories for traumatic events.[68] While it is becoming increasingly clear that epigenetic mechanisms likely regulate gene transcription during memory consolidation, very few of the known epigenetic regulators have actually been identified to contribute to the reconsolidation and extinction processes. As a result, our understanding of the epigenetic regulation of memory reconsolidation and extinction is only its early stages. However, considering that histone acetylation can both promote and inhibit memory retention using reconsolidation-based strategies, it is critical to have a better understanding of not only how histone acetylation regulates the reconsolidation process but also how it interacts with other epigenetic mechanisms to properly modify memories. Future studies will likely aim to address the epigenetic regulators of transcriptional regulation in the reconsolidation and extinction processes in more detail.

Epigenetic mechanisms are a unique way by which gene transcription can be transiently and persistently regulated without altering the genetic code. Considering that memory formation and storage require transient changes in gene expression that are thought to regulate permanent changes in synaptic strength, posttranslational modification of histones and methylation of DNA could potentially act as an important regulator of short-term and long-term changes in gene expression necessary to maintain the memory trace at the molecular level. Ultimately, epigenetic regulation of *de novo* gene transcription may be a dynamic regulator of memory formation, storage, modification, and persistence throughout the life span.

ACKNOWLEDGMENTS

This work was supported in part by the National Institutes of Health grant MH097909 (F. D. L.) and the Evelyn F. McKnight Brain Institute at the University of Alabama at Birmingham. The authors declare no competing financial interests.

REFERENCES

1. Maren S, Phan KL, Liberzon I. The contextual brain: implications for fear conditioning, extinction and psychopathology. *Nat Rev Neurosci*. 2013;14:417–428.
2. LeDoux JE. Emotion circuits in the brain. *Annu Rev Neurosci*. 2000;23:155–184.
3. Gal-Ben-Ari S, Rosenblum K. Molecular mechanisms underlying memory consolidation of taste information in the cortex. *Front Behav Neurosci*. 2011;5:87.
4. Pavlov IP. *Conditioned Reflexes: An Investigation of the Physiological Activity of the Cerebral Cortex*. London: Oxford University Press; 1927.
5. Jarome TJ, Helmstetter FJ. The ubiquitin–proteasome system as a critical regulator of synaptic plasticity and long-term memory formation. *Neurobiol Learn Mem*. 2013;105:107–116.
6. McGaugh JL. Memory—a century of consolidation. *Science*. 2000;287:248–251.
7. Nader K, Schafe GE, Le Doux JE. Fear memories require protein synthesis in the amygdala for reconsolidation after retrieval. *Nature*. 2000;406:722–726.
8. Alberini CM. The role of reconsolidation and the dynamic process of long-term memory formation and storage. *Front Behav Neurosci*. 2011;5:12.
9. Diaz-Mataix L, Ruiz Martinez RC, Schafe GE, Ledoux JE, Doyere V. Detection of a temporal error triggers reconsolidation of amygdala-dependent memories. *Curr Biol*. 2013;23:467–472.
10. Monfils MH, Cowansage KK, Klann E, LeDoux JE. Extinction-reconsolidation boundaries: key to persistent attenuation of fear memories. *Science*. 2009;324:951–955.
11. Myers KM, Davis M. Mechanisms of fear extinction. *Mol Psychiatry*. 2007;12:120–150.
12. Parsons RG, Ressler KJ. Implications of memory modulation for post-traumatic stress and fear disorders. *Nat Neurosci*. 2013;16:146–153.
13. Davis HP, Squire LR. Protein synthesis and memory: a review. *Psychol Bull*. 1984;96:518–559.
14. Johansen JP, Cain CK, Ostroff LE, LeDoux JE. Molecular mechanisms of fear learning and memory. *Cell*. 2011;147:509–524.
15. Gupta-Agarwal S, Franklin AV, Deramus T, et al. G9a/GLP histone lysine dimethyltransferase complex activity in the hippocampus and the entorhinal cortex is required for gene activation and silencing during memory consolidation. *J Neurosci*. 2012;32:5440–5453.
16. Kerimoglu C, Agis-Balboa RC, Kranz A, et al. Histone-methyltransferase MLL2 (KMT2B) is required for memory formation in mice. *J Neurosci*. 2013;33:3452–3464.
17. Alberini CM. Transcription factors in long-term memory and synaptic plasticity. *Physiol Rev*. 2009;89:121–145.

18. Silva AJ, Kogan JH, Frankland PW, Kida S. CREB and memory. *Annu Rev Neurosci.* 1998;21:127–148.
19. Swank MW, Sweatt JD. Increased histone acetyltransferase and lysine acetyltransferase activity and biphasic activation of the ERK/RSK cascade in insular cortex during novel taste learning. *J Neurosci.* 2001;21:3383–3391.
20. Zovkic IB, Guzman-Karlsson MC, Sweatt JD. Epigenetic regulation of memory formation and maintenance. *Learn Mem.* 2013;20:61–74.
21. Marmorstein R. Protein modules that manipulate histone tails for chromatin regulation. *Nat Rev Mol Cell Biol.* 2001;2:422–432.
22. Lubin FD, Gupta S, Parrish RR, Grissom NM, Davis RL. Epigenetic mechanisms: critical contributors to long-term memory formation. *Neuroscientist.* 2011;17(6): 616–632.
23. Guan JS, Haggarty SJ, Giacometti E, et al. HDAC2 negatively regulates memory formation and synaptic plasticity. *Nature.* 2009;459:55–60.
24. McQuown SC, Wood MA. HDAC3 and the molecular brake pad hypothesis. *Neurobiol Learn Mem.* 2011;96:27–34.
25. Stefanko DP, Barrett RM, Ly AR, Reolon GK, Wood MA. Modulation of long-term memory for object recognition via HDAC inhibition. *Proc Natl Acad Sci USA.* 2009;106:9447–9452.
26. Glozak MA, Sengupta N, Zhang X, Seto E. Acetylation and deacetylation of non-histone proteins. *Gene.* 2005;363:15–23.
27. Jarome TJ, Lubin FD. Histone lysine methylation: critical regulator of memory and behavior. *Rev Neurosci.* 2013;24:375–387.
28. Gupta S, Kim SY, Artis S, et al. Histone methylation regulates memory formation. *J Neurosci.* 2010;30:3589–3599.
29. Chwang WB, O'Riordan KJ, Levenson JM, Sweatt JD. ERK/MAPK regulates hippocampal histone phosphorylation following contextual fear conditioning. *Learn Mem.* 2006;13:322–328.
30. Atkins CM, Selcher JC, Petraitis JJ, Trzaskos JM, Sweatt JD. The MAPK cascade is required for mammalian associative learning. *Nat Neurosci.* 1998;1:602–609.
31. Rodrigues SM, Schafe GE, LeDoux JE. Intra-amygdala blockade of the NR2B subunit of the NMDA receptor disrupts the acquisition but not the expression of fear conditioning. *J Neurosci.* 2001;21:6889–6896.
32. Chwang WB, Arthur JS, Schumacher A, Sweatt JD. The nuclear kinase mitogen- and stress-activated protein kinase 1 regulates hippocampal chromatin remodeling in memory formation. *J Neurosci.* 2007;27:12732–12742.
33. Jarome TJ, Werner CT, Kwapis JL, Helmstetter FJ. Activity dependent protein degradation is critical for the formation and stability of fear memory in the amygdala. *PLoS One.* 2011;6:e24349.
34. Ma MK, Heath C, Hair A, West AG. Histone crosstalk directed by H2B ubiquitination is required for chromatin boundary integrity. *PLoS Genet.* 2011;7:e1002175.
35. Castro-Gomez S, Barrera-Ocampo A, Machado-Rodriguez G, et al. Specific de-SUMOylation triggered by acquisition of spatial learning is related to epigenetic changes in the rat hippocampus. *Neuroreport.* 2013;24:976–981.
36. Day JJ, Sweatt JD. Cognitive neuroepigenetics: a role for epigenetic mechanisms in learning and memory. *Neurobiol Learn Mem.* 2011;96:2–12.
37. Feng J, Zhou Y, Campbell SL, et al. Dnmt1 and Dnmt3a maintain DNA methylation and regulate synaptic function in adult forebrain neurons. *Nat Neurosci.* 2010;13:423–430.
38. Miller CA, Sweatt JD. Covalent modification of DNA regulates memory formation. *Neuron.* 2007;53:857–869.

39. Lubin FD, Roth TL, Sweatt JD. Epigenetic regulation of BDNF gene transcription in the consolidation of fear memory. *J Neurosci*. 2008;28:10576–10586.
40. Kaas GA, Zhong C, Eason DE, et al. TET1 controls CNS 5-methylcytosine hydroxylation, active DNA demethylation, gene transcription, and memory formation. *Neuron*. 2013;79:1086–1093.
41. Miller CA, Gavin CF, White JA, et al. Cortical DNA methylation maintains remote memory. *Nat Neurosci*. 2010;13:664–666.
42. Kwapis JL, Jarome TJ, Lonergan ME, Helmstetter FJ. Protein kinase Mzeta maintains fear memory in the amygdala but not in the hippocampus. *Behav Neurosci*. 2009;123:844–850.
43. Migues PV, Hardt O, Wu DC, et al. PKMzeta maintains memories by regulating GluR2-dependent AMPA receptor trafficking. *Nat Neurosci*. 2010;13:630–634.
44. Pastalkova E, Serrano P, Pinkhasova D, Wallace E, Fenton AA, Sacktor TC. Storage of spatial information by the maintenance mechanism of LTP. *Science*. 2006;313: 1141–1144.
45. Serrano P, Friedman EL, Kenney J, et al. PKMzeta maintains spatial, instrumental, and classically conditioned long-term memories. *PLoS Biol*. 2008;6:2698–2706.
46. Lubin FD. Epigenetic gene regulation in the adult mammalian brain: multiple roles in memory formation. *Neurobiol Learn Mem*. 2011;96:68–78.
47. Puckett RE, Lubin FD. Epigenetic mechanisms in experience-driven memory formation and behavior. *Epigenomics*. 2011;3:649–664.
48. Alberini CM, Ledoux JE. Memory reconsolidation. *Curr Biol*. 2013;23:R746–R750.
49. Lubin FD, Sweatt JD. The IkappaB kinase regulates chromatin structure during reconsolidation of conditioned fear memories. *Neuron*. 2007;55:942–957.
50. Si J, Yang J, Xue L, et al. Activation of NF-kappaB in basolateral amygdala is required for memory reconsolidation in auditory fear conditioning. *PLoS One*. 2012;7:e43973.
51. Bredy TW, Barad M. The histone deacetylase inhibitor valproic acid enhances acquisition, extinction, and reconsolidation of conditioned fear. *Learn Mem*. 2008;15:39–45.
52. Maddox SA, Watts CS, Schafe GE. p300/CBP histone acetyltransferase activity is required for newly acquired and reactivated fear memories in the lateral amygdala. *Learn Mem*. 2013;20:109–119.
53. Han JH, Kushner SA, Yiu AP, et al. Selective erasure of a fear memory. *Science*. 2009;323:1492–1496.
54. Lee SH, Choi JH, Lee N, et al. Synaptic protein degradation underlies destabilization of retrieved fear memory. *Science*. 2008;319:1253–1256.
55. Kaang BK, Choi JH. Synaptic protein degradation in memory reorganization. *Adv Exp Med Biol*. 2012;970:221–240.
56. Maddox SA, Schafe GE. Epigenetic alterations in the lateral amygdala are required for reconsolidation of a Pavlovian fear memory. *Learn Mem*. 2011;18:579–593.
57. Maddox SA, Watts CS, Schafe GE. DNA methyltransferase activity is required for memory-related neural plasticity in the lateral amygdala. *Neurobiol Learn Mem*. 2014;107:93–100.
58. Schiller D, Monfils MH, Raio CM, Johnson DC, Ledoux JE, Phelps EA. Preventing the return of fear in humans using reconsolidation update mechanisms. *Nature*. 2010;463:49–53.
59. Graff J, Joseph NF, Horn ME, et al. Epigenetic priming of memory updating during reconsolidation to attenuate remote fear memories. *Cell*. 2014;156:261–276.
60. Kim JJ, Fanselow MS. Modality-specific retrograde amnesia of fear. *Science*. 1992;256:675–677.
61. Clem RL, Huganir RL. Calcium-permeable AMPA receptor dynamics mediate fear memory erasure. *Science*. 2010;330:1108–1112.

62. Bredy TW, Wu H, Crego C, Zellhoefer J, Sun YE, Barad M. Histone modifications around individual BDNF gene promoters in prefrontal cortex are associated with extinction of conditioned fear. *Learn Mem*. 2007;14:268–276.
63. Lattal KM, Wood MA. Epigenetics and persistent memory: implications for reconsolidation and silent extinction beyond the zero. *Nat Neurosci*. 2013;16:124–129.
64. Marek R, Coelho CM, Sullivan RK, et al. Paradoxical enhancement of fear extinction memory and synaptic plasticity by inhibition of the histone acetyltransferase p300. *J Neurosci*. 2011;31:7486–7491.
65. Wei W, Coelho CM, Li X, et al. p300/CBP-associated factor selectively regulates the extinction of conditioned fear. *J Neurosci*. 2012;32:11930–11941.
66. Rudenko A, Dawlaty MM, Seo J, et al. Tet1 is critical for neuronal activity-regulated gene expression and memory extinction. *Neuron*. 2013;79:1109–1122.
67. Miller CA, Campbell SL, Sweatt JD. DNA methylation and histone acetylation work in concert to regulate memory formation and synaptic plasticity. *Neurobiol Learn Mem*. 2008;89:599–603.
68. Alberini CM, Chen DY. Memory enhancement: consolidation, reconsolidation and insulin-like growth factor 2. *Trends Neurosci*. 2012;35:274–283.

CHAPTER TWO

Epigenetics in Posttraumatic Stress Disorder

Carina Rampp*, Elisabeth B. Binder*,†, Nadine Provençal*
*Department of Translational Research in Psychiatry, Max Planck Institute of Psychiatry, Munich, Germany
†Department of Psychiatry and Behavioral Sciences, Emory University School of Medicine, Atlanta, Georgia, USA

Contents

Abstract

Reported exposure to traumatic event is relatively common within the general population (40–90%), but only a fraction of individuals will develop posttraumatic stress disorder (PTSD). Indeed, the lifetime prevalence of PTSD is estimated to range between 7% and 12%. The factors influencing risk or resilience to PTSD after exposure to traumatic events are likely both environmental, such as type, timing, and extent of trauma, and genetic. Recently, epigenetic mechanisms have been implicated in mediating altered risk for PTSD as they can reflect both genetic and environmental influences. In this chapter, we describe the accumulating evidences for epigenetic factors in PTSD highlighting the importance of sensitive periods as well as methodological aspects such as tissue availabilities for such studies. We describe studies using a candidate gene approach focusing mainly on key players in the stress hormone regulation that show epigenetic alterations both in humans and in animal models for PTSD. We also summarize the results of epigenome-wide studies reporting associations with PTSD. For the above, we focus on one epigenetic mechanism, DNA methylation, as it is so far the best studied for this disorder. Finally, we describe how epigenetic mechanisms could be responsible for the long-lasting effects of gene–environment interactions observed in PTSD.

Progress in Molecular Biology and Translational Science, Volume 128
ISSN 1877-1173
http://dx.doi.org/10.1016/B978-0-12-800977-2.00002-4

1. INTRODUCTION

Posttraumatic stress disorder (PTSD) is a psychiatric condition characterized by persistent symptoms of intrusive reexperiencing, avoidance, and autonomous–vegetative hyperarousal following exposure to a traumatic event. With a prevalence of around 5% in the general population and a lifetime prevalence ranging between 7% and 12%, it belongs to the most frequent psychiatric disorders whereat it is twice as common in women than in men.[1] PTSD goes along with extensive suffering not only for the affected individual but also on a family and society level and constitutes a major burden for the health system. A traumatic event that can lead to PTSD is defined as an event that represents an intense threat for the own life, health, or physical integrity and goes along with feelings of strong fear and helplessness. Immediate witnessing of such an event can also lead to traumatization in the absence of personal victimization. PTSD can occur soon after the traumatic event or with delay. Dependent on the definition of trauma and the observed sample, a lifetime trauma incidence ranging between 40% and 90% has been reported in the general population.[2,3] These high trauma exposure rates together with a relatively smaller prevalence of PTSD in the population indicate that trauma exposure does not necessarily lead to the development of PTSD. As generally observed in psychiatric disorders, trauma exposure has a stronger impact in more vulnerable individuals with a higher probability to develop a pathological phenotype.[4,5] Indeed, different risk and resilience factors have been discussed for PTSD.[4,6]

One major risk factor that has been identified in PTSD is the type of the trauma itself. It could be shown, for example, that traumatic events caused directly by other individuals like physical violence or sexual assault involve a greater risk for PTSD than impersonal catastrophes like natural disasters.[1,7] Here, among others, the amount of social support might play an important role, since it has been identified as a decisive protective factor in processing a traumatic experience.[8] Often, the latter type of trauma affects many individuals at once and thus is accompanied by stronger social support, whereas the former mostly affects single individuals. In addition, the time of trauma during life plays a decisive role. Until two decades ago, the hypothesis that early-life trauma is associated with an increased risk of adult mood and anxiety disorders was supported largely by anecdotal reports inspired by psychoanalytic concepts of early critical periods of development.[9] During the past years, an increasing amount of studies explored the consequences of early trauma on psychosocial and psychobiological development and clearly showed that the childhood period is a very sensitive period and that adverse experiences during that time coincide with a significantly higher risk for PTSD[10–12] as well as other psychiatric disorders[13–15] later in life. These observations are increasingly confirmed by neurobiological research that identified possible mechanisms by which early-life trauma can have long-lasting impact on pathophysiology of PTSD.[16]

The exploration of neurobiological and molecular principles underlying the susceptibility for PTSD and other affective disorders is currently a matter of intense research. The research for genetic variants mediating vulnerability for affective and anxiety disorders showed only limited success with no strong main genetic effects reported[6,17] as is apparent from large genome-wide association studies for major depressive disorder (MDD).[18–20] It is now well established that psychiatric phenotypes such as PTSD emerge from a complex interplay of genetic and environmental factors as well as epigenetic factors as it could be shown more recently.[6] Epigenetic mechanisms, such as DNA methylation, not only program cell identity but also have been shown to play an important role in the biological response triggered by environmental changes.[21] In recent years, the dynamic role of epigenetic regulation in response to the environment has become clear. In adaption to a stressor, an organism reacts with epigenetic changes that lead to changes in gene expression profiles and in consequence to alterations of biological pathways and of neuronal functioning.[16] If the stressor overstrains capacities of an organism to adapt, it can lead to maladaptive changes and dysregulation of biological systems such as the hypothalamic–pituitary–adrenal (HPA) axis. Early trauma can engrave biological marks and exert a life-long impact on health and disease trajectories leading to a higher risk not only for psychiatric but also for internal diseases and an impaired immune function.[22,23]

This chapter will describe the accumulating evidences for epigenetic alterations in PTSD highlighting the importance of sensitive periods as well as methodological aspects such as tissue availabilities. It will highlight studies that used a candidate gene approach focusing mainly on HPA axis key players as well as epigenome-wide studies reporting DNA methylation alterations in humans and animal models for PTSD. Finally, it will also shed light onto how epigenetic mechanisms could be responsible for the long-lasting effects of gene–environment interactions observed in PTSD.

2. EPIGENETIC EVIDENCES IN PTSD

In addition to the DNA sequence that has been shown to regulate gene expression as well as to influence the risk for PTSD independently of the type of trauma exposure,[24,25] epigenetic information is another layer of transcriptional regulation.[26] The main functions of the epigenome are to regulate gene transcription and to compact the DNA into the cell nucleus. Several distinct epigenetic marks come together to achieve this, including DNA methylation and hydroxymethylation, histone modifications, ATP-dependent chromatin remodeling, and noncoding RNAs. These marks can prime not only current but also future gene expression and protein translation modifications making it a prime target to understand the effect of early trauma on the development of PTSD later in life.

DNA methylation is a covalent modification of the cytosine residues that are located primarily but not exclusively at CpG dinucleotide sequences in mammals.[27,28] Increased DNA methylation in the promoter region of a gene is usually associated with repressed gene expression[29] as it can interfere with the binding of transcriptional enhancer complexes.[30] Intragenic DNA methylation is also observed,[27,31–33] and in contrast to promoter methylation, it is found to correlate both negatively and positively with gene expression[34–37] as it can also interfere with the binding of transcriptional repressive complexes.[38,39] To accurately maintain the DNA methylation profiles and prevent a drift in the DNA methylation pattern during the life course, several biochemical elements come into play. DNA methylation patterns are copied and maintained by DNA methyltransferases (DNMTs).[40] Removal of methyl groups or DNA demethylation is achieved either through direct action of demethylases[41] or through DNA excision/repair-based mechanisms[42–47] involving intermediate steps, such as hydroxymethylation.

As mentioned in Section 1, there is a growing body of evidence suggesting that in addition to the innate endogenous processes sculpting the DNA methylation pattern during gestation, the DNA methylation pattern is responsive to external environmental exposures in animals[48–50] and in humans.[16,51–53] This would allow to prime future responses of an organism depending on its early environment.[21,54,55]

Using a candidate gene approach, researchers have identified DNA methylation alterations in brain tissues following diverse postnatal stresses in various genes such as the *glucocorticoid receptor* (*GR*),[56] *vasopressin* (*AVP*),[57] and *brain-derived neurotrophic factor*[58] genes in animals, as well as

GR[59] and *ribosomal RNA*[60] genes in humans. In these studies, early-life stress and DNA methylation alterations were associated with various behavioral abnormalities in adulthood such as increase stress reactivity and suicide. Whole genomic approaches in humans and animals have also been used and have identified numerous additional genes and functional pathways to be altered following child abuse.[61–63]

Several mechanisms have been proposed as to how environmental exposures lead to alterations in brain DNA methylation profiles. One is the dissociation of the methyl-binding protein, MeCP2, following its phosphorylation induced by neuronal activation (depolarization)[64,65] that leads to demethylation and increased expression of the *AVP* gene in the paraventricular nucleus in mice exposed to maternal deprivation.[57] In addition, Weaver *et al.* identified that maternal licking and grooming in rats triggered DNA demethylation at the *GR* locus through the activation of serotonin signaling pathways that increases transcription factor (nerve growth factor-inducible A) binding and subsequent *GR* expression.[56,66] Rats exposed to lower levels of maternal care displayed higher levels of DNA methylation and lower *GR* expression associated with increase stress reactivity. Furthermore, it was shown that the binding of activated GR to the glucocorticoid response elements (GREs) located in gene regulatory regions can lead to demethylation of CpGs around the site and, by that, changes in genome-wide transcriptional regulation.[67,68] Although these studies mostly investigated specific gene targets, they strongly indicate that early-life environment has an impact on DNA methylation in the brain that can be mediated through specific signaling pathways following neuronal activation.[53]

2.1. Methodological aspects and tissue specificity

In contrast to animal studies and a limited number of human studies mentioned above, most of the clinical and epidemiological work currently ongoing does not include brain samples. Thus, epigenetic studies have used peripheral tissue such as blood and buccal cells and identified numerous epigenetic associations with early-life adversity in the context of diseases. These associations can be seen as simple biomarkers but might also be causally involved in the disease state through their action on the HPA axis and immune system, for example.[69]

In a recent study described in detail below, we analyzed gene expression and DNA methylation profiles in peripheral blood cells among individuals with PTSD, with or without exposure to severe child abuse and identified distinct patterns of gene expression and DNA methylation profiles associated with child abuse.[70] Suderman *et al.* also identified a specific DNA methylation signature in peripheral blood DNA associated with childhood abuse.[71] Less severe childhood experiences such as lower socioeconomic position in childhood but not in adulthood were also shown to associate with specific gene expression[72] and DNA methylation[73] profiles in adult peripheral tissue. Not only blood cells but also buccal epithelial cells DNA methylation alterations have been found to associate with parental stress in adolescents[74] and victimization/bullying in discordant monozygotic twins.[75]

Although different signatures in response to early-life events are observed in different tissues, epigenetic changes in peripheral tissues may correlate to some extent with measures in the brain. As expected, Davies *et al.* observed that between-tissue variation in DNA methylation greatly exceeds between-individual differences within any one tissue in healthy controls. However, they also identified selected interindividual variations that were indeed correlated across both brain and blood samples.[76] Furthermore, differential DNA methylation of the *serotonin transporter* (*SLC6A4*) gene promoter in blood cells observed with childhood-limited physical aggression in men was also found associated with *in vivo* measures of human brain serotonin synthesis.[77] These data indicate that peripheral tissues could be useful in clinical and epidemiological studies of complex neurobiological phenotypes.

The observations that DNA methylation changes associated with early adversity are not limited to the brain but are also observed in peripheral systems highlight the fact that such changes may be induced by system-wide influences. The long-lasting effects of aversive environmental exposures are thus likely to be reflected in tissues outside the brain with probably additional health risks. In rhesus macaques, the broad impact of maternal deprivation on DNA methylation profiles was seen in both the brain and T cells, supporting the hypothesis that the response to early-life adversity is system-wide and genome-wide and persists to adulthood.[78]

DNA methylation studies in PTSD have therefore used peripheral and neural DNA samples combined with both gene candidates and genomic approaches to delineate the involvement of epigenetic mechanisms in the development of the disease in humans and animal models.

2.2. Candidate gene approach in animals and humans

2.2.1 Focus on the HPA axis genes

The HPA axis is the most important system in stress regulation. The hypothalamus as superior center of the axis produces among others the hormones corticotropin-releasing hormone (CRH) and arginine AVP, which act in a synergistic and complementary way in stress reaction. These hormones stimulate the distribution of adrenocorticotropic hormone (ACTH) from the pituitary gland. ACTH again triggers the release of sexual hormones, aldosterone, and cortisol from the adrenal glands. Cortisol acts together with catecholamines as primary effector of the stress system. Cortisol actions are mediated through its binding to the GR and the mineralocorticoid receptor (MR)[79] that act as transcription factors[79,80] and are expressed in most tissues. Binding of cortisol to the GR and MR induces their translocation into the nucleus where they can exert their function as transcription factors regulating adaptive responses to stress, including metabolism, immune activation, and cell proliferation and differentiation. To regain homeostasis after a stressor, the HPA axis regulates itself via a negative feedback mechanism mediated via neuronal GRs. Therefore, a decrease in GR expression/activation is generally associated with an increase in the response to stress due to an impaired negative feedback.

Dysregulation of the HPA axis is one of the most robust findings in biological psychiatry.[81] In MDD, for example, a GR hyposensitivity has been described leading to an impaired negative feedback mechanism via neuronal GRs and consecutively to increased release of stress hormones resulting in a hyperactivity of the HPA axis. In contrast to the findings in MDD, in PTSD, most studies observed a hypersensitivity of the GR accompanied by a super-suppression of cortisol following low-dose dexamethasone, increased CRH levels, and lower baseline cortisol levels.[6,82] Since the HPA axis is one of the main systems activated after exposure to a trauma and perturbations in this system have been frequently described in PTSD, genes regulating this system are prime candidates for epigenetic PTSD research.

Most studies on DNA methylation patterns in humans examine either patients with PTSD or healthy individuals with a trauma history. While a number of studies have implicated epigenetic changes following early trauma exposure in genes regulating the HPA axis, such as the *GR* gene (*NR3C1*) itself[83,84] and the genes encoding the hypothalamic peptides in CRH[85] and AVP[57,85] (see above), fewer studies investigated these changes in these genes in PTSD.

Most of the human studies focused on the *NR3C1* gene. Labonte *et al.* compared cortisol, GR expression, and promoter methylation levels in blood cells of 16 healthy controls without trauma exposure versus 30 individuals with a diagnosis of lifetime PTSD (current or remitted). Individuals with lifetime PTSD showed lower morning cortisol release, higher GR expression, and lower overall methylation levels in the examined GR promoter regions. GR expression was inversely correlated with cortisol levels and with methylation levels. No difference between current PTSD and remitted PTSD was found.[86] Consistent with these findings, Yehuda *et al.* reported significantly lower GR promoter methylation levels in veterans with PTSD compared to those without PTSD in a sample of 122 combat-exposed veterans. GR promoter methylation levels were inversely correlated with cortisol decline in the dexamethasone suppression test and with PTSD-related symptoms.[87] Yehuda *et al.* also showed that methylation levels may be changed by psychotherapy and may serve as a predictor for treatment outcome. The *NR3C1* and the *FKBP5* gene (a cochaperone of the GR that is described in detail in section 3) were examined in a sample of 16 combat veterans with PTSD who received prolonged exposure psychotherapy. The authors observed that methylation of the GR promoter assessed before treatment predicted treatment outcome, but was not significantly altered in responders or nonresponders at posttreatment or follow-up. In contrast, methylation of the FKBP5 promoter region did not predict treatment response, but showed variation in association with treatment outcome with decreased levels of promoter methylation among responders in the posttreatment period. Psychotherapy may thus be seen as an environmental regulator of epigenetic states.[88] There are also studies that examined the molecular biological background of transgenerational transmission of vulnerability for PTSD focusing on GR promoter methylation and considering neuroendocrine measures. Perroud *et al.* found that mothers exposed to the Tutsi genocide during pregnancy as well as their children showed higher methylation levels in GR and MR promoter regions compared to nonexposed mothers and their children. In the neuroendocrine measures, the exposed mothers and their children had lower cortisol and GR levels and MR levels.[89] These results do not agree with those in studies reported above, where PTSD was associated with lower methylation levels and GR supersensitivity. This underlines inconsistencies in the current literature that necessitate further exploration. Among other factors, ethnicity and type and timing of the traumatic events might play a role. Another example of transgenerational epigenetic effect is presented by Yehuda *et al.* who reported differential effects of maternal versus paternal PTSD on epigenetic regulation and neuroendocrine activity in the offspring of Holocaust survivors. In the absence of maternal PTSD, offspring with paternal PTSD showed higher GR-1_F promoter methylation, whereas offspring with both maternal PTSD and paternal PTSD showed lower methylation. Again, lower GR promoter methylation was significantly associated with greater cortisol suppression. The moderation of paternal PTSD effects by maternal PTSD suggests different mechanisms for the intergenerational transmission of trauma-related vulnerabilities depending on the parental lineage.[90]

2.2.2 Other stress-responsive genes

Other stress-responsive genes have also been explored in epigenetic research on PTSD. The genes encoding for pituitary adenylate cyclase-activating polypeptide (PACAP) and its receptor (PAC1) have been shown to be associated with PTSD in a sex-specific manner.[91] Among many other functions like cell signaling, metabolism, homeostasis, and cell protection, PACAP has been shown to broadly regulate the cellular stress response. In a highly traumatized sample consisting of subjects with or without PTSD, Ressler *et al.* examined the association of the PACAP–PAC1 pathway with PTSD. They analyzed blood levels of PACAP, genetic variation, and methylation of the PACAP (*ADCYAP1*) and PAC1 receptor (*ADCYAP1R1*) genes. They found PACAP levels to be associated with PTSD diagnosis and symptoms in females. After analyzing 44 SNPs in the two genes, a single SNP in a putative estrogen response element within *ADCYAP1R1* turned out to predict PTSD diagnosis and symptoms in females and was associated with *ADCYAP1R1* mRNA expression level. Further analyses showed that methylation within the first CpG island of *ADCYAP1R1* was positively correlated with PTSD symptoms in both sexes with higher methylation levels in affected subjects. For the gene encoding the peptide itself, *ADCYAP1*, no significant association with PTSD was found. Complementing these human data, rodent models showed that *ADCYAP1R1* mRNA is induced with fear conditioning or estrogen replacement. Together, the data suggest that perturbations in the PACAP/PAC1 pathway are involved in abnormal stress responses, where *ADCYAP1R1* gene regulation possibly occurs under both genetic and epigenetic mechanisms in a sex-specific manner.[91]

Next to genes regulating the stress response, also genes regulating neurotransmitter balance are first-line candidates in PTSD research. Several studies reported a correlation between methylation of the serotonin transporter gene, *SLC6A4*, and PTSD to traumatic events. A study with 100 participants, 23 of them suffering from PTSD, could not detect a direct association between *SLC6A4* methylation status and PTSD, but found a modifying influence of *SLC6A4* methylation levels on the effect of the number of traumatic events on PTSD. Persons with more traumatic events were at increased risk for PTSD, but only at lower methylation levels.

At higher methylation levels, individuals with more traumatic events were protected from this disorder.[92] Most methylation studies investigating *SLC6A4* did not use PTSD as outcome phenotype, but rather looked at the association of methylation patterns with traumatic events. van IJzendoorn *et al.* examined 143 adopted participants for psychological problems due to unresolved loss or trauma. They found that associations between *SLC6A4* polymorphisms (5HTTLPR) and psychological problems were significantly altered by environmentally induced methylation patterns.[93] Other studies on *SLC6A4* gene found an association of CpG methylation levels with childhood abuse.[94,95]

Also, the dopamine system has been of scientific interest in PTSD. In a sample of 62 PTSD patients and 258 controls, Chang *et al.* found a significant association of a polymorphism in the dopamine transporter gene *SLC6A3* with PTSD. 9R allele carriers had almost double the risk of lifetime PTSD compared to 10R/10R genotype carriers. In a subset of 16 PTSD patients and 67 controls, they could show that 9R allele carriers were only at higher risk for PTSD in conjunction with high methylation in the *SLC6A3* promoter locus.[96] In addition, the *catechol-O-methyl transferase* (*COMT*) gene encoding for the COMT enzyme, which is responsible for the depletion of catecholamines like dopamine, has also been reported to associate with PTSD.[97] It could be shown that higher methylation at CpG sites in the promoter region of this gene was associated with impaired fear inhibition in PTSD.[98]

Most of these findings suggest that epigenetic mechanisms are associated with the development of PTSD in many genes where prior work also revealed a genetic risk factor. Together, they suggest gene–environment interactions with epigenetic factors acting as an interface modulating the effect of adverse life events on the organism.

2.3. Genome-wide studies in animals and humans

Up to now, most methylation studies focused on candidate genes. Genome-wide studies however offer a less biased approach to identify prime targets and have been used to explore the effect of traumatic events and early environmental influences on DNA methylation in PTSD patients as well as in trauma-exposed healthy individuals.

In a genome-wide study with 169 participants, we were able to show that PTSD patients had different gene expression and DNA methylation profiles in their blood cells, dependent on the presence of childhood maltreatment.[70]

To do so, we compared two PTSD groups, one with childhood abuse ($n = 32$) and one without childhood abuse ($n = 29$), to a trauma-exposed control group ($n = 108$). The control group participants were free of PTSD in their past medical history and did not show current PTSD symptoms. Compared to the control group, PTSD cases without child abuse showed 244 differentially regulated transcripts, while PTSD cases with child abuse showed 303 differentially regulated transcripts. Surprisingly, the overlap of differentially expressed transcripts between the two PTSD groups was only 2%, indicating that distinct biomarker profiles in PTSD occur dependent on the presence or absence of child abuse.[70] Moreover, these gene expression differences were correlated to DNA methylation changes. For this purpose, DNA methylation profiles of the transcripts differentially regulated in either of the two PTSD groups versus controls were investigated. DNA methylation changes underlying the observed gene expression differences were 2- to 12-fold higher in PTSD patients with child abuse compared to those without child abuse. These results indicate that changes in gene expression are more often associated with changes in DNA methylation in PTSD occurring after child abuse than PTSD occurring in the absence of child abuse. A subsequently conducted pathway analysis showed several overlapping networks between the two PTSD groups mainly concerning cell function and immunity. This overlap despite the completely divergent transcripts between the two groups suggests a possible common downstream effect of disease in blood cells. On the other hand, several nonoverlapping networks were observed like the enrichment of the central nervous system development and tolerance induction pathways in the PTSD group with child abuse and the apoptosis and growth rate networks in the PTSD group without child abuse. Together, these data suggest that differential biological mechanisms might be responsible for the observed gene expression differences in PTSD depending on the time and severity of the adverse environmental exposure.

These data are consistent with the accumulating evidence showing that the epigenome is highly responsive to adverse experiences during childhood, whereas environmental influences during adulthood have a less permanent effect on the epigenome and consequently act through different mechanisms that are less stable in the development of diseases.

As mentioned above, trauma-induced epigenetic changes may have system-wide influences on the organism and on health trajectories. A reciprocal interference of the HPA axis and the immune system has been described suggesting an association of PTSD with alterations in immune function.[99] Indeed, PTSD was shown to be associated with DNA methylation changes in blood cells in genes that are involved in immune function.[100] In a study with 100 participants (23 PTSD-affected and 77 nonaffected individuals), Uddin *et al.* examined genome-wide methylation profiles for over 14,000 genes. The number of uniquely methylated genes was significantly higher in PTSD-affected patients. Moreover, they found a distinct signature reflecting immune activation among PTSD-affected individuals. According to their findings, exposure to a potentially traumatic event induces changes in methylation profiles in vulnerable persons resulting in gene expression changes accompanied by altered immune function.[100] These findings emphasize the system-wide influence of epigenetic changes and the major role of the immune system in PTSD.

3. EPIGENETIC MECHANISMS IN GENE BY ENVIRONMENT INTERACTIONS IN PTSD

As shown in the preceding paragraphs, there is growing scientific evidence for the long-term epigenetic effects of early environment, little is known about how this is moderated by genetic variation though. In the past few years, it has become clear that genetic and environmental factors interact in a complex and multiplicative—rather than additive way—in the development of multifactorial disorders like PTSD.[101–103] These so-called gene × environment interactions are increasingly considered in psychiatric research. It is likely that this approach comes closer to the complex genesis of stress-related disorders than looking at individual factors. The genetic makeup of an organism and its individual interplay of the genetic code with epigenetic modulations seem to play a major role in the susceptibility to stress-related disorders.

Up to now, most GxE studies in PTSD focused on candidate genes. Since the HPA axis is responsive to external stressors and mediates their effects on a biological and also gene expression level, genes regulating this system represent major candidates also for GxE research. One of these is the *FKBP5* gene, which modulates GR sensitivity and thus HPA regulation. A functional haplotype in this gene showed a significant interaction with severity of child abuse to predict adult PTSD symptoms. Carriers of the risk alleles were at higher risk to develop PTSD in adulthood in the presence of a childhood trauma.[104] We could show a genotype-dependent DNA methylation interacting with early trauma in the *FKBP5* gene as a potential molecular mechanism for GxE in PTSD.[105]

The investigated gene in this study, *FKBP5*, plays a central role in the regulation of the HPA axis. FKBP5 acts as an inhibitor of the GR reducing its ability to bind glucocorticoids and its translocation into the nucleus. By reducing the GR's sensitivity to glucocorticoids, increased FKBP5 levels can lead to an impaired negative feedback mechanism and thus to a dysregulation of the HPA axis and a prolonged stress response.[106,107] This endocrine feature has previously shown to be associated with depression and exposure to early trauma.[13] FKBP5 expression is induced following stress exposure through the binding of the GR to GREs located in several regulatory regions of the gene.[108] This is part of an ultrashort intracellular negative feedback loop for GR activity.[109] Risk allele carriers of the upper mentioned functional haplotype show significantly higher glucocorticoid-induced FKBP5 expression accompanied by higher GR resistance in healthy individuals. This constellation presumably leads to a higher vulnerability for different psychiatric diseases.

A tag SNP of the described haplotype in the *FKBP5* gene (rs1360780) was found to influence the three-dimensional loop formation between an intronic glucocorticoid responsive enhancer in intron 2 and the transcription start site of the gene following its transcriptional activation by stress. This intronic SNP, located close to a functional GRE, alters not only the molecular response to stress but also the systemic response by leading to prolonged cortisol release after stress[110,111] and altered amygdala and hippocampal activation in threat-related tasks.[112–114] With exposure to trauma in childhood, the initial deregulation of FKBP5 response to stress by the genetic predisposition is likely followed by glucocorticoid-induced DNA demethylation of another enhancer element in intron 7 of the gene. Indeed, we saw that the risk allele promotes a 3D DNA conformation that leads to a local proximity of the GRE in intron 2 to a functional GRE in intron 7 and to the transcription start site. This conformation favors CpG demethylation after GR binding in the functionally relevant GRE in intron 7 in the risk allele carriers resulting in increased FKBP5 transcription. This genotype-dependent demethylation and gene expression are also associated with differences in hippocampal volume and immune transcription.[105] Environmentally induced epigenetic changes may thus have sequence-specific effects.

Polymorphisms in the FKBP5 gene have also been shown to be associated with a row of psychiatric phenotypes in addition to PTSD such as MDD, suicidal behavior, aggression, and psychotic symptoms in Alzheimer's disease.[115] While in MDD, T-allele carriers show reduced

GR sensitivity, the opposite has been observed in PTSD. Interestingly, the GR supersensitivity in PTSD is genotype-dependent and only present in carriers of the FKBP5 risk alleles. This state-dependent GR supersensitivity could be found in several studies.[104,110,116,117] The reasons for these disease- and genotype-dependent findings likely involve distinct epigenetic changes that need to be understood using longitudinal studies, dissecting the effects of trauma exposure and subsequent development of the disease.

4. CONCLUSIONS

A number of studies have investigated and reported epigenetic changes in PTSD, on both the candidate and genome-wide level. The changes could relate to the long-term effects of trauma that are either adaptive or maladaptive but could also reflect preexisting epigenetic changes that alter the impact of trauma on the organism. Indeed, a significant part of the PTSD symptoms observed in war zone veterans is explained by preexisting factors such as predeployment psychological symptoms or prior life stressors.[118] Early-life stress such as child abuse is an important risk factor for PTSD and is also known to cause long-lasting epigenetic alterations that further increase the risk for PTSD. In addition, these epigenetic factors will likely interact with genetic factors, such as in the case of *FKBP5*,[105] which can alter the susceptibility of the individual to the adverse effects of trauma exposure. Also, these effects on risk versus resilience to pathophysiology of such complex interactions will likely change across the different developmental stages of the organism affecting, for example, their HPA axis regulation at different levels through life. To disentangle these differences, longitudinal studies with the integration of multiple measures will be necessary in PTSD research.[10]

Epigenetic studies in PTSD so far have identified alterations in DNA methylation profiles using mainly peripheral DNA samples as blood and saliva are accessible tissues in human cohorts. As mentioned earlier, both tissue- and cell-specific changes will be apparent in the context of fear conditioning as well as system-wide effects, which are most likely seen in response to stress hormones.[16] Indeed, the increased release of glucocorticoids observed following early-life stress could simultaneously target similar genes across diverse tissues since its receptors, GRs, are present in most tissues.[68] Subsequently, DNA methylation changes could be induced in several tissues by these activated hormones and receptors in response to early-life stress.[119,120] For example, the GxE interaction associated with methylation changes in the stress response regulator gene, *FKBP5*, described earlier in blood cells, leads to similar DNA methylation changes after glucocorticoid exposure in human progenitor hippocampal neuronal cells.[105] In addition, Mehta *et al.* had revealed that brain relevant pathways are enriched among genes with early trauma-related differences in DNA methylation in peripheral blood cells.[70] These results have important practical implications for utilization of peripheral blood tissues for identification of biomarkers in PTSD but could also help us in the understanding of the pathophysiological mechanisms of disease since an increasing number of studies point to an immune component in PTSD.[121–123]

Since epigenetic alterations are potentially reversible by either drugs or behavioral therapy such as demonstrated in maternal care studies in animals[55,124] and possibly humans,[90] this opens up new promising therapeutic avenues for PTSD. It has been proposed that the epigenetic mechanisms involved in normal fear learning and memory in animals also extend to the development of persistent and pathological fear characteristics observed in PTSD.[125] Inhibitors of DNA-methylating enzymes (DNMTs) have been shown to disrupt synaptic plasticity such as long-term potentiation (LTP)[126,127] that is also known to be a mediator of learned fear in PTSD.[128,129] Interestingly, these studies demonstrated that these effects on LTP could be reversed by histone-modifying enzymes such as histone deacetylases.[126,127] Together, these data indicate that research focusing on an in-depth understanding of the epigenetic players in PTSD is much needed and will possibly lead to new pharmacological interventions that could be applied not only in PTSD treatment but also and more importantly in the prevention of PTSD such as in the context of conditioned fear-extinction models.[125]

REFERENCES

1. Breslau N, Kessler RC, Chilcoat HD, et al. Trauma and posttraumatic stress disorder in the community: the 1996 Detroit Area Survey of Trauma. *Arch Gen Psychiatry.* 1998;55:626–632.
2. Breslau N. The epidemiology of posttraumatic stress disorder: what is the extent of the problem? *J Clin Psychiatry.* 2001;62(suppl 17):16–22.
3. Kessler RC, Sonnega A, Bromet E, Hughes M, Nelson CB. Posttraumatic stress disorder in the National Comorbidity Survey. *Arch Gen Psychiatry.* 1995;52:1048–1060.
4. Gillespie CF, Phifer J, Bradley B, Ressler KJ. Risk and resilience: genetic and environmental influences on development of the stress response. *Depress Anxiety.* 2009;26:984–992.
5. Booij L, Wang D, Levesque ML, Tremblay RE, Szyf M. Looking beyond the DNA sequence: the relevance of DNA methylation processes for the stress-diathesis model of depression. *Philos Trans R Soc Lond B Biol Sci.* 2013;368:20120251.

6. Mehta D, Binder EB. Gene × environment vulnerability factors for PTSD: the HPA-axis. *Neuropharmacology*. 2012;62:654–662.
7. Prigerson HG, Maciejewski PK, Rosenheck RA. Combat trauma: trauma with highest risk of delayed onset and unresolved posttraumatic stress disorder symptoms, unemployment, and abuse among men. *J Nerv Ment Dis*. 2001;189:99–108.
8. Galea S, Resnick H, Ahern J, et al. Posttraumatic stress disorder in Manhattan, New York City, after the September 11th terrorist attacks. *J Urban Health*. 2002;79:340–353.
9. Nemeroff CB, Bremner JD, Foa EB, et al. Posttraumatic stress disorder: a state-of-the-science review. *J Psychiatr Res*. 2006;40:1–21.
10. Yehuda R, Flory JD, Pratchett LC, et al. Putative biological mechanisms for the association between early life adversity and the subsequent development of PTSD. *Psychopharmacology (Berl)*. 2010;212:405–417.
11. Widom CS. Posttraumatic stress disorder in abused and neglected children grown up. *Am J Psychiatry*. 1999;156:1223–1229.
12. Bremner JD, Southwick SM, Johnson DR, Yehuda R, Charney DS. Childhood physical abuse and combat-related posttraumatic stress disorder in Vietnam veterans. *Am J Psychiatry*. 1993;150:235–239.
13. Heim C, Nemeroff CB. The role of childhood trauma in the neurobiology of mood and anxiety disorders: preclinical and clinical studies. *Biol Psychiatry*. 2001;49: 1023–1039.
14. Widom CS. The cycle of violence. *Science*. 1989;244:160–166.
15. Rutter M. Developmental catch-up, and deficit, following adoption after severe global early privation. English and Romanian Adoptees (ERA) Study Team. *J Child Psychol Psychiatry*. 1998;39:465–476.
16. Klengel T, Pape J, Binder EB, Mehta D. The role of DNA methylation in stress-related psychiatric disorders. *Neuropharmacology*. 2014;80:115–132.
17. Klengel T, Binder EB. Gene-environment interactions in major depressive disorder. *Can J Psychiatry*. 2013;58:76–83.
18. Wray NR, Pergadia ML, Blackwood DH, et al. Genome-wide association study of major depressive disorder: new results, meta-analysis, and lessons learned. *Mol Psychiatry*. 2012;17:36–48.
19. Hek K, Demirkan A, Lahti J, et al. A genome-wide association study of depressive symptoms. *Biol Psychiatry*. 2013;73:667–678.
20. Ripke S, O'Dushlaine C, Chambert K, et al. Genome-wide association analysis identifies 13 new risk loci for schizophrenia. *Nat Genet*. 2013;45:1150–1159.
21. Meaney MJ. Epigenetics and the biological definition of gene × environment interactions. *Child Dev*. 2010;81:41–79.
22. Widom CS, DuMont K, Czaja SJ. A prospective investigation of major depressive disorder and comorbidity in abused and neglected children grown up. *Arch Gen Psychiatry*. 2007;64:49–56.
23. Danese A, Moffitt TE, Harrington H, et al. Adverse childhood experiences and adult risk factors for age-related disease: depression, inflammation, and clustering of metabolic risk markers. *Arch Pediatr Adolesc Med*. 2009;163:1135–1143.
24. Stein MB, Jang KL, Taylor S, Vernon PA, Livesley WJ. Genetic and environmental influences on trauma exposure and posttraumatic stress disorder symptoms: a twin study. *Am J Psychiatry*. 2002;159:1675–1681.
25. True WR, Rice J, Eisen SA, et al. A twin study of genetic and environmental contributions to liability for posttraumatic stress symptoms. *Arch Gen Psychiatry*. 1993;50:257–264.
26. Razin A. CpG methylation, chromatin structure and gene silencing—a three-way connection. *EMBO J*. 1998;17:4905–4908.

27. Lister R, Pelizzola M, Dowen RH, et al. Human DNA methylomes at base resolution show widespread epigenomic differences. *Nature*. 2009;462:315–322.
28. Xie W, Barr CL, Kim A, et al. Base-resolution analyses of sequence and parent-of-origin dependent DNA methylation in the mouse genome. *Cell*. 2012;148:816–831.
29. Bird AP. CpG-rich islands and the function of DNA methylation. *Nature*. 1986;321:209–213.
30. Murgatroyd C, Spengler D. Histone tales: echoes from the past, prospects for the future. *Genome Biol*. 2010;11:105.
31. Ball MP, Li JB, Gao Y, et al. Targeted and genome-scale strategies reveal gene-body methylation signatures in human cells. *Nat Biotechnol*. 2009;27:361–368.
32. Maunakea AK, Chepelev I, Zhao K. Epigenome mapping in normal and disease states. *Circ Res*. 2010;107:327–339.
33. Ogoshi K, Hashimoto S, Nakatani Y, et al. Genome-wide profiling of DNA methylation in human cancer cells. *Genomics*. 2011;98:280–287.
34. Jiang Y, Liu S, Chen X, Cao Y, Tao Y. Genome-wide distribution of DNA methylation and DNA demethylation and related chromatin regulators in cancer. *Biochim Biophys Acta*. 2013;1835:155–163.
35. Jjingo D, Conley AB, Yi SV, Lunyak VV, Jordan IK. On the presence and role of human gene-body DNA methylation. *Oncotarget*. 2012;3:462–474.
36. Mehta D, Raison CL, Woolwine BJ, et al. Transcriptional signatures related to glucose and lipid metabolism predict treatment response to the tumor necrosis factor antagonist infliximab in patients with treatment-resistant depression. *Brain Behav Immun*. 2013;31:205–215.
37. Provencal N, Suderman MJ, Caramaschi D, et al. Differential DNA methylation regions in cytokine and transcription factor genomic loci associate with childhood physical aggression. *PLoS One*. 2013;8:e71691.
38. Ando T, Nishimura M, Oka Y. Decitabine (5-aza-2′-deoxycytidine) decreased DNA methylation and expression of MDR-1 gene in K562/ADM cells. *Leukemia*. 2000;14:1915–1920.
39. Bartolomei MS. Genomic imprinting: employing and avoiding epigenetic processes. *Genes Dev*. 2009;23:2124–2133.
40. Jeltsch A. Reading and writing DNA methylation. *Nat Struct Mol Biol*. 2008;15: 1003–1004.
41. Ramchandani S, Bhattacharya SK, Cervoni N, Szyf M. DNA methylation is a reversible biological signal. *Proc Natl Acad Sci USA*. 1999;96:6107–6112.
42. Barreto G, Schafer A, Marhold J, et al. Gadd45a promotes epigenetic gene activation by repair-mediated DNA demethylation. *Nature*. 2007;445:671–675.
43. Ma DK, Guo JU, Ming GL, Song H. DNA excision repair proteins and Gadd45 as molecular players for active DNA demethylation. *Cell Cycle*. 2009;8:1526–1531.
44. Morgan HD, Dean W, Coker HA, Reik W, Petersen-Mahrt SK. Activation-induced cytidine deaminase deaminates 5-methylcytosine in DNA and is expressed in pluripotent tissues: implications for epigenetic reprogramming. *J Biol Chem*. 2004;279: 52353–52360.
45. Rai K, Huggins IJ, James SR, et al. DNA demethylation in zebrafish involves the coupling of a deaminase, a glycosylase, and gadd45. *Cell*. 2008;135:1201–1212.
46. Schmitz KM, Schmitt N, Hoffmann-Rohrer U, et al. TAF12 recruits Gadd45a and the nucleotide excision repair complex to the promoter of rRNA genes leading to active DNA demethylation. *Mol Cell*. 2009;33:344–353.
47. Guo JU, Su Y, Zhong C, Ming GL, Song H. Hydroxylation of 5-methylcytosine by TET1 promotes active DNA demethylation in the adult brain. *Cell*. 2011;145: 423–434.

48. Darnaudery M, Maccari S. Epigenetic programming of the stress response in male and female rats by prenatal restraint stress. *Brain Res Rev*. 2008;57:571–585.
49. Gudsnuk K, Champagne FA. Epigenetic influence of stress and the social environment. *ILAR J*. 2012;53:279–288.
50. Jirtle RL, Skinner MK. Environmental epigenomics and disease susceptibility. *Nat Rev Genet*. 2007;8:253–262.
51. Mill J, Tang T, Kaminsky Z, et al. Epigenomic profiling reveals DNA-methylation changes associated with major psychosis. *Am J Hum Genet*. 2008;82:696–711.
52. Sasaki A, de Vega WC, McGowan PO. Biological embedding in mental health: an epigenomic perspective. *Biochem Cell Biol*. 2013;91:14–21.
53. Szyf M. The early-life social environment and DNA methylation. *Clin Genet*. 2012;81:341–349.
54. Mill J, Petronis A. Pre- and peri-natal environmental risks for attention-deficit hyperactivity disorder (ADHD): the potential role of epigenetic processes in mediating susceptibility. *J Child Psychol Psychiatry*. 2008;49:1020–1030.
55. Szyf M. DNA methylation, the early-life social environment and behavioral disorders. *J Neurodev Disord*. 2011;3:238–249.
56. Weaver IC, Diorio J, Seckl JR, Szyf M, Meaney MJ. Early environmental regulation of hippocampal glucocorticoid receptor gene expression: characterization of intracellular mediators and potential genomic target sites. *Ann NY Acad Sci*. 2004;1024:182–212.
57. Murgatroyd C, Patchev AV, Wu Y, et al. Dynamic DNA methylation programs persistent adverse effects of early-life stress. *Nat Neurosci*. 2009;12:1559–1566.
58. Roth TL, Lubin FD, Funk AJ, Sweatt JD. Lasting epigenetic influence of early-life adversity on the BDNF gene. *Biol Psychiatry*. 2009;65:760–769.
59. McGowan PO, Sasaki A, D'Alessio AC, et al. Epigenetic regulation of the glucocorticoid receptor in human brain associates with childhood abuse. *Nat Neurosci*. 2009;12:342–348.
60. McGowan PO, Sasaki A, Huang TC, et al. Promoter-wide hypermethylation of the ribosomal RNA gene promoter in the suicide brain. *PLoS One*. 2008;3:e2085.
61. Weder N, Zhang H, Jensen K, et al. Child abuse, depression, and methylation in genes involved with stress, neural plasticity, and brain circuitry. *J Am Acad Child Adolesc Psychiatry*. 2014;53(4):417–424.e5. http://dx.doi.org/10.1016/j.jaac.2013.12.025. Epub 2014 Jan 27.
62. Suderman M, McGowan PO, Sasaki A, et al. Conserved epigenetic sensitivity to early life experience in the rat and human hippocampus. *Proc Natl Acad Sci USA*. 2012;109(suppl 2):17266–17272.
63. Labonte B, Suderman M, Maussion G, et al. Genome-wide epigenetic regulation by early-life trauma. *Arch Gen Psychiatry*. 2012;69:722–731.
64. Chen Y, Sharples TJ, Phillips KG, et al. The nicotinic alpha 4 beta 2 receptor selective agonist, TC-2559, increases dopamine neuronal activity in the ventral tegmental area of rat midbrain slices. *Neuropharmacology*. 2003;45:334–344.
65. Martinowich K, Hattori D, Wu H, et al. DNA methylation-related chromatin remodeling in activity-dependent BDNF gene regulation. *Science*. 2003;302:890–893.
66. Weaver IC, D'Alessio AC, Brown SE, et al. The transcription factor nerve growth factor-inducible protein a mediates epigenetic programming: altering epigenetic marks by immediate-early genes. *J Neurosci*. 2007;27:1756–1768.
67. Kress C, Thomassin H, Grange T. Active cytosine demethylation triggered by a nuclear receptor involves DNA strand breaks. *Proc Natl Acad Sci USA*. 2006;103:11112–11117.
68. Thomassin H, Flavin M, Espinas ML, Grange T. Glucocorticoid-induced DNA demethylation and gene memory during development. *EMBO J*. 2001;20:1974–1983.

69. Miller DB, O'Callaghan JP. Aging, stress and the hippocampus. *Ageing Res Rev.* 2005;4:123–140.
70. Mehta D, Klengel T, Conneely KN, et al. Childhood maltreatment is associated with distinct genomic and epigenetic profiles in posttraumatic stress disorder. *Proc Natl Acad Sci USA.* 2013;110:8302–8307.
71. Suderman M, Borghol N, Pappas JJ, et al. Childhood abuse is associated with methylation of multiple loci in adult DNA. *BMC Med Genomics.* 2014;7:13.
72. Miller GE, Chen E, Fok AK, et al. Low early-life social class leaves a biological residue manifested by decreased glucocorticoid and increased proinflammatory signaling. *Proc Natl Acad Sci USA.* 2009;106:14716–14721.
73. Borghol N, Suderman M, McArdle W, et al. Associations with early-life socio-economic position in adult DNA methylation. *Int J Epidemiol.* 2012;41:62–74.
74. Essex MJ, Thomas Boyce W, Hertzman C, et al. Epigenetic vestiges of early developmental adversity: childhood stress exposure and DNA methylation in adolescence. *Child Dev.* 2013;84(1):58–75. http://dx.doi.org/10.1111/j.1467-8624.2011.01641.x. Epub 2011 Sep 2.
75. Ouellet-Morin I, Wong CC, Danese A, et al. Increased serotonin transporter gene (SERT) DNA methylation is associated with bullying victimization and blunted cortisol response to stress in childhood: a longitudinal study of discordant monozygotic twins. *Psychol Med.* 2013;43:1813–1823.
76. Davies MN, Volta M, Pidsley R, et al. Functional annotation of the human brain methylome identifies tissue-specific epigenetic variation across brain and blood. *Genome Biol.* 2012;13:R43.
77. Wang D, Szyf M, Benkelfat C, et al. Peripheral SLC6A4 DNA methylation is associated with in vivo measures of human brain serotonin synthesis and childhood physical aggression. *PLoS One.* 2012;7:e39501.
78. Provencal N, Suderman MJ, Guillemin C, et al. The signature of maternal rearing in the methylome in rhesus macaque prefrontal cortex and T cells. *J Neurosci.* 2012;32: 15626–15642.
79. Larsson L, Thorbert-Mros S, Rymo L, Berglundh T. Influence of epigenetic modifications of the interleukin-10 promoter on IL10 gene expression. *Eur J Oral Sci.* 2012;120:14–20.
80. Shen P, Xie ZJ, Li H, Sanchez ER. Glucocorticoid receptor conversion to high affinity nuclear binding and transcription enhancement activity in Chinese hamster ovary cells subjected to heat and chemical stress. *J Steroid Biochem Mol Biol.* 1993;47:55–64.
81. Holsboer F. The corticosteroid receptor hypothesis of depression. *Neuropsychopharmacology.* 2000;23:477–501.
82. Yehuda R. Status of glucocorticoid alterations in post-traumatic stress disorder. *Ann NY Acad Sci.* 2009;1179:56–69.
83. Perroud N, Paoloni-Giacobino A, Prada P, et al. Increased methylation of glucocorticoid receptor gene (NR3C1) in adults with a history of childhood maltreatment: a link with the severity and type of trauma. *Transl Psychiatry.* 2011;1:e59.
84. Tyrka AR, Price LH, Marsit C, Walters OC, Carpenter LL. Childhood adversity and epigenetic modulation of the leukocyte glucocorticoid receptor: preliminary findings in healthy adults. *PLoS One.* 2012;7:e30148.
85. Elliott E, Ezra-Nevo G, Regev L, Neufeld-Cohen A, Chen A. Resilience to social stress coincides with functional DNA methylation of the Crf gene in adult mice. *Nat Neurosci.* 2010;13:1351–1353.
86. Labonte B, Azoulay N, Yerko V, Turecki G, Brunet A. Epigenetic modulation of glucocorticoid receptors in posttraumatic stress disorder. *Transl Psychiatry.* 2014;4:e368.

87. Yehuda R, Flory JD, Bierer LM, et al. Lower methylation of glucocorticoid receptor gene promoter 1 in peripheral blood of veterans with posttraumatic stress disorder. *Biol Psychiatry*. 2014, in press. pii: S0006-3223(14)00100-0. http://dx.doi.org/10.1016/j.biopsych.2014.02.006. [Epub ahead of print].
88. Yehuda R, Daskalakis NP, Desarnaud F, et al. Epigenetic biomarkers as predictors and correlates of symptom improvement following psychotherapy in combat veterans with PTSD. *Front Psychiatry*. 2013;4:118.
89. Perroud N, Rutembesa E, Paoloni-Giacobino A, et al. The Tutsi genocide and transgenerational transmission of maternal stress: epigenetics and biology of the HPA axis. *World J Biol Psychiatry*. 2014;15:334–345.
90. Yehuda R, Daskalakis NP, Lehrner A, et al. Influences of maternal and paternal PTSD on epigenetic regulation of the glucocorticoid receptor gene in holocaust survivor offspring. *Am J Psychiatry*. 2014;171(8):872–880.
91. Ressler KJ, Mercer KB, Bradley B, et al. Post-traumatic stress disorder is associated with PACAP and the PAC1 receptor. *Nature*. 2011;470:492–497.
92. Koenen KC, Uddin M, Chang SC, et al. SLC6A4 methylation modifies the effect of the number of traumatic events on risk for posttraumatic stress disorder. *Depress Anxiety*. 2011;28:639–647.
93. van IMH, Caspers K, Bakermans-Kranenburg MJ, Beach SR, Philibert R. Methylation matters: interaction between methylation density and serotonin transporter genotype predicts unresolved loss or trauma. *Biol Psychiatry*. 2010;68:405–407.
94. Beach SR, Brody GH, Todorov AA, Gunter TD, Philibert RA. Methylation at 5HTT mediates the impact of child sex abuse on women's antisocial behavior: an examination of the Iowa adoptee sample. *Psychosom Med*. 2011;73:83–87.
95. Vijayendran M, Beach SR, Plume JM, Brody GH, Philibert RA. Effects of genotype and child abuse on DNA methylation and gene expression at the serotonin transporter. *Front Psychiatry*. 2012;3:55.
96. Chang SC, Koenen KC, Galea S, et al. Molecular variation at the SLC6A3 locus predicts lifetime risk of PTSD in the Detroit Neighborhood Health Study. *PLoS One*. 2012;7:e39184.
97. Clark R, DeYoung CG, Sponheim SR, et al. Predicting post-traumatic stress disorder in veterans: interaction of traumatic load with COMT gene variation. *J Psychiatr Res*. 2013;47:1849–1856.
98. Norrholm SD, Jovanovic T, Smith AK, et al. Differential genetic and epigenetic regulation of catechol-O-methyltransferase is associated with impaired fear inhibition in posttraumatic stress disorder. *Front Behav Neurosci*. 2013;7:30.
99. Wong CM. Post-traumatic stress disorder: advances in psychoneuroimmunology. *Psychiatr Clin North Am*. 2002;25:369–383, vii.
100. Uddin M, Aiello AE, Wildman DE, et al. Epigenetic and immune function profiles associated with posttraumatic stress disorder. *Proc Natl Acad Sci USA*. 2010;107:9470–9475.
101. Newton-Cheh C, Hirschhorn JN. Genetic association studies of complex traits: design and analysis issues. *Mutat Res*. 2005;573:54–69.
102. Clarke MC, Tanskanen A, Huttunen M, Whittaker JC, Cannon M. Evidence for an interaction between familial liability and prenatal exposure to infection in the causation of schizophrenia. *Am J Psychiatry*. 2009;166:1025–1030.
103. van Os J, Rutten BP, Poulton R. Gene-environment interactions in schizophrenia: review of epidemiological findings and future directions. *Schizophr Bull*. 2008;34:1066–1082.

104. Binder EB, Bradley RG, Liu W, et al. Association of FKBP5 polymorphisms and childhood abuse with risk of posttraumatic stress disorder symptoms in adults. *JAMA*. 2008;299:1291–1305.
105. Klengel T, Mehta D, Anacker C, et al. Allele-specific FKBP5 DNA demethylation mediates gene-childhood trauma interactions. *Nat Neurosci*. 2013;16:33–41.
106. Binder EB, Salyakina D, Lichtner P, et al. Polymorphisms in FKBP5 are associated with increased recurrence of depressive episodes and rapid response to antidepressant treatment. *Nat Genet*. 2004;36:1319–1325.
107. Binder EB. The role of FKBP5, a co-chaperone of the glucocorticoid receptor in the pathogenesis and therapy of affective and anxiety disorders. *Psychoneuroendocrinology*. 2009;34(suppl 1):S186–S195.
108. Jaaskelainen T, Makkonen H, Palvimo JJ. Steroid up-regulation of FKBP51 and its role in hormone signaling. *Curr Opin Pharmacol*. 2011;11:326–331.
109. Vermeer H, Hendriks-Stegeman BI, van der Burg B, van Buul-Offers SC, Jansen M. Glucocorticoid-induced increase in lymphocytic FKBP51 messenger ribonucleic acid expression: a potential marker for glucocorticoid sensitivity, potency, and bioavailability. *J Clin Endocrinol Metab*. 2003;88:277–284.
110. Ising M, Depping AM, Siebertz A, et al. Polymorphisms in the FKBP5 gene region modulate recovery from psychosocial stress in healthy controls. *Eur J Neurosci*. 2008;28:389–398.
111. Buchmann AF, Holz N, Boecker R, et al. Moderating role of FKBP5 genotype in the impact of childhood adversity on cortisol stress response during adulthood. *Eur Neuropsychopharmacol*. 2014;24:837–845.
112. Fani N, Gutman D, Tone EB, et al. FKBP5 and attention bias for threat: associations with hippocampal function and shape. *JAMA Psychiatry*. 2013;70:392–400.
113. Holz NE, Buchmann AF, Boecker R, et al. Role of FKBP5 in emotion processing: results on amygdala activity, connectivity and volume. *Brain Struct Funct*. 2014, in press.
114. White MG, Bogdan R, Fisher PM, et al. FKBP5 and emotional neglect interact to predict individual differences in amygdala reactivity. *Genes Brain Behav*. 2012;11:869–878.
115. Zannas AS, Binder EB. Gene-environment interactions at the FKBP5 locus: sensitive periods, mechanisms and pleiotropism. *Genes Brain Behav*. 2014;13:25–37.
116. Menke A, Klengel T, Rubel J, et al. Genetic variation in FKBP5 associated with the extent of stress hormone dysregulation in major depression. *Genes Brain Behav*. 2013;12:289–296.
117. Mehta D, Gonik M, Klengel T, et al. Using polymorphisms in FKBP5 to define biologically distinct subtypes of posttraumatic stress disorder: evidence from endocrine and gene expression studies. *Arch Gen Psychiatry*. 2011;68:901–910.
118. Franz MR, Wolf EJ, MacDonald HZ, et al. Relationships among predeployment risk factors, warzone-threat appraisal, and postdeployment PTSD symptoms. *J Trauma Stress*. 2013;26:498–506.
119. Szyf M. Early life, the epigenome and human health. *Acta Paediatr*. 2009;98:1082–1084.
120. Szyf M, Bick J. DNA methylation: a mechanism for embedding early life experiences in the genome. *Child Dev*. 2013;84:49–57.
121. Uddin M, Aiello AE, Wildman DE, et al. Epigenetic and immune function profiles associated with posttraumatic stress disorder. *Proc Natl Acad Sci USA*. 2011;107:9470–9475.
122. Pace TW, Wingenfeld K, Schmidt I, et al. Increased peripheral NF-kappaB pathway activity in women with childhood abuse-related posttraumatic stress disorder. *Brain Behav Immun*. 2012;26:13–17.

123. Spitzer C, Wibisono D, Terfehr K, et al. C-reactive protein, pre- and post-dexamethasone cortisol levels in post-traumatic stress disorder. *Nord J Psychiatry*. 2014;68:296–299.
124. Szyf M, Weaver I, Meaney M. Maternal care, the epigenome and phenotypic differences in behavior. *Reprod Toxicol*. 2007;24:9–19.
125. Zovkic IB, Sweatt JD. Epigenetic mechanisms in learned fear: implications for PTSD. *Neuropsychopharmacology*. 2013;38:77–93.
126. Levenson JM, Roth TL, Lubin FD, et al. Evidence that DNA (cytosine-5) methyltransferase regulates synaptic plasticity in the hippocampus. *J Biol Chem*. 2006;281: 15763–15773.
127. Monsey MS, Ota KT, Akingbade IF, Hong ES, Schafe GE. Epigenetic alterations are critical for fear memory consolidation and synaptic plasticity in the lateral amygdala. *PLoS One*. 2011;6:e19958.
128. Rogan MT, Staubli UV, LeDoux JE. Fear conditioning induces associative long-term potentiation in the amygdala. *Nature*. 1997;390:604–607.
129. Koshibu K, Graff J, Mansuy IM. Nuclear protein phosphatase-1: an epigenetic regulator of fear memory and amygdala long-term potentiation. *Neuroscience*. 2011;173:30–36.

CHAPTER THREE

Histone-Mediated Epigenetics in Addiction

Leah N. Hitchcock, K. Matthew Lattal
Department of Behavioral Neuroscience, Oregon Health & Science University, Portland, Oregon, USA

Contents

Abstract

Many of the brain regions, neurotransmitter systems, and behavioral changes that occur after occasional drug use in healthy subjects and after chronic drug abuse in addicted patients are well characterized. An emerging literature suggests that epigenetic processes, those processes that regulate the accessibility of DNA to regulatory proteins

Progress in Molecular Biology and Translational Science, Volume 128
ISSN 1877-1173
http://dx.doi.org/10.1016/B978-0-12-800977-2.00003-6

within the nucleus, are keys to how addiction develops and how it may be treated. Investigations of the regulation of chromatin, the organizational system of DNA, by histone modification are leading to a new understanding of the cellular and behavioral alterations that occur after drug use. We will describe how, when, and where histone tails are modified and how some of the most recognized histone regulation patterns are involved in the cycle of addiction, including initial and chronic drug intake, withdrawal, abstinence, and relapse. Finally, we consider how an approach that targets histone modifications may promote successful treatment.

1. INTRODUCTION

Adapting to a constantly changing environment requires the ability to acquire new behaviors and change old ones in response to environmental contingencies. For this reason, much of our behavior is controlled by its consequences, with response probabilities in given situations changing as a function of the history of reinforcement in those situations. These experiences are learned and remembered through several well-defined neural circuits that are involved in acquisition, consolidation, and expression of memories. In addiction, however, the reinforcing properties of the abused substance can be so powerful that the plasticity mechanisms involved in normal learning and memory are usurped, resulting in repetitive and persistent behaviors that are resistant to change, even in the face of strong negative consequences. One key to understanding addiction is to understand the basic mechanisms of learning and memory and how those mechanisms may be altered by abused substances (e.g., Ref. 1). A great deal is now known about the ways in which memories are formed, from binding neurotransmitters at the receptor level to the activation of transcriptional machinery needed for the synthesis of new proteins that solidify long-term memories.

A growing literature indicates that long-term memories form at a molecular level as a consequence of changes in gene expression induced by activity-dependent histone modifications (e.g., Refs. 2,3). These same mechanisms are involved in acquiring and stabilizing the long-term reinforcing effects of various drugs of abuse. In this chapter, we examine the relation between histone modifications and long-term memory in behavioral approaches that model different aspects of addiction. We consider how these modifications may interact with the behavioral experience (acquisition and retrieval of memories for abused substances and extinction of drug-seeking behavior) and review some different theoretical perspectives that have been offered for these effects. We begin this chapter with a brief overview of epigenetics and the general memory processes that may occur in the development and maintenance of addiction. We then describe different histone-mediated mechanisms that may control long-term gene expression and plasticity and review current literature demonstrating how these mechanisms may be involved in different addiction processes. Finally, we end this chapter with a description of currently unresolved issues in the field of histone-mediated epigenetics in addiction, and we suggest several future directions that may help to resolve some of these debates.

2. GENERAL OVERVIEW OF HISTONE-MEDIATED EPIGENETICS

The term epigenetics refers to the regulation that occurs "epi" or "over" the genomic DNA. Although controversy about the term epigenetics remains (to be discussed later in the chapter), most research supports the idea that chemical modifications applied to the DNA or near DNA (i.e., to histone proteins) help regulate transcription and are modifiable throughout life. Within the nucleus of a cell, DNA is wrapped around multiple nucleosomes. Nucleosomes are composed of and linked together by a collection of histone proteins. Histones are the small and positively charged building blocks that help package and organize DNA into a repeat bead-like structure. To allow for selective and modifiable outcomes through epigenetic processes, each histone is classified into one of two superfamilies (i.e., histone core or histone linker), five families (H2A, H2B, H3, H4, and H1/H5), and multiple subfamilies, each having slightly different functions and cellular distribution patterns.[4,5] Two copies of each histone core (H2A, H2B, H3, and H4) are bound together by linker histones (H1/H5) to create one nucleosome (an octamer of core histones) for DNA to be carefully wrapped around, making the basic chromatin structure (i.e., nucleosome + DNA). In this chapter, we will consider how these histones may be modified during different stages of drug taking—acute exposure, chronic drug taking, withdrawal, abstinence, and relapse. Before going into histone alterations in greater detail, we will consider addiction at the level of behavior and how different learning and memory processes contribute to long-term drug intake.

3. LEARNING PROCESSES INVOLVED IN THE DEVELOPMENT OF ADDICTION

One of the reasons that addiction is thought to involve learning and memory circuits is that drug seeking often occurs in the presence of specific cues. These cues can be contextual (e.g., drinking a beer in a favorite bar), social (e.g., using cocaine with a specific group of friends), and temporal (e.g., having a cigarette first thing in the morning), among others. Over time, drug seeking occurs in different situations, broadening the cues that may be associated with drug intake. After even a relatively few number of experiences, those cues will evoke powerful drug cravings when they are encountered. These cravings create a negative internal state that further motivates drug intake. From a learning and memory perspective, the challenge is understanding how these cues become associated with drugs and what can be done to sever or, at the very least, suppress those powerful associations.

Basic research on learning and memory processes in substance abuse has focused on mechanisms that underlie two very general learning processes: initial acquisition, in which the memory is initially formed and consolidated, and long-term maintenance, in which the memory is retrieved and modulated. Research on epigenetic mechanisms in memory aspects of addiction has largely focused on simple memory processes—how cue—drug associations are initially encoded, consolidated, and retrieved. This research has revealed the critical importance of histone acetylation and gene expression in mediating several aspects of these memory processes. We begin our review of histone-mediated epigenetics in addiction with a description of some of these memory processes in more detail.

3.1. Initial establishment of drug-associated memories

When patterns of drug use first begin, new associations are encoded between the drug and the user's environment, consolidated into a memory, and later retrieved when cues associated with drug seeking are encountered. The very first experience with a drug of abuse activates circuits that are involved in reward and in learning and memory. With this first exposure, the initial memory begins to form. This memory likely involves distal contexts and discrete cues associated with some aspect of the drug of abuse. Theoretical approaches to memory have found that upon this initial exposure to the drug, the memory is labile for a period of time before it is stabilized through a time-limited consolidation process.[6] In some cases, a single experience with a drug of abuse can lead to a lasting memory; in other cases, this memory is established and strengthened more incrementally, with repeated exposures to the drug increasing the strength of the memory and distributing the representation of that memory across different circuits in the brain. As these memories are being established, they can be modified by additional processes that are triggered by drugs of abuse, such as sensitization, tolerance, and withdrawal.[7,8] Together, these processes ultimately result in habitual drug seeking that results from an interaction of circuits mediating contextual information (e.g., the hippocampus), response initiation and maintenance (e.g., the striatum), and reward value (e.g., nucleus accumbens (NAc) and amygdala).

These initial memory processes contribute to the development of the repeated binge/intoxication stage of drug addiction, driven by a collection of brain regions in an excitatory circuit. The first time drugs of abuse are used (e.g., psychostimulants, alcohol, opioids, nicotine, and Δ9-tetrahydrocannabinol), brain regions in both reward (i.e., ventral tegmental area (VTA), striatum, NAc core, and thalamus) and learning (basolateral and central nucleus of the amygdala, medial prefrontal cortex, and hippocampus) centers share excitatory information (reviewed in Refs. 9,10). Some of these structures serve a primary role in particular stages of addiction (e.g., binge/intoxication, withdrawal/negative affect, and preoccupation/anticipation/craving[9]) and memory processes (i.e., encoding, consolidation, and retrieval; e.g., Refs. 11,12), yet many of these regions can be recruited throughout each process. Importantly, the circuits mediating aspects of reward and aspects of memory overlap, with key processes in the amygdala, VTA, and NAc controlling consolidation of these drug memories.

Current research on histone-mediated epigenetics in addiction focuses on the role of histone acetylation specifically in memory consolidation. Broadly speaking, this histone modification may create a permissive state in chromatin, promoting access by transcription factors (TFs) to the genomic DNA, thereby facilitating gene expression. Histone modifications alone do not cause long-term memory—they must be paired with activation of signaling cascades that are triggered by exposure to drugs of abuse. Once formed, these memories can be modulated in several different ways when additional exposure to the drug occurs in specific environments.

3.2. Retrieval of drug-associated memories and extinction of drug-seeking behavior

Once memories between environmental cues and drugs of abuse are established, there are several consequences that occur with subsequent exposure to those cues and drugs. First, these memories may evoke cravings, causing an increase in drug-seeking behavior and drug consumption (e.g., Ref. 13). Repeated cravings and drug administration result in tolerance and withdrawal, two processes that are key to maintaining addiction.[14]

Second, the act of retrieval will trigger some of the initial processes—encoding and consolidation of the retrieved memory—and will potentially include new contextual components that were absent during initial acquisition. This additional consolidation may simply recapitulate the initial consolidation process (i.e., the idea of reconsolidation[15]), and, in addition, they almost certainly involve consolidation of new memories specific to this new experience (e.g., Ref. 16). These consolidation and reconsolidation processes are thought to contribute to the long-term maintenance of addiction. Third, if the drug that is expected (based on the retrieval of a previous memory) is not consumed, extinction may begin to develop. If no drug is administered during repeated retrieval episodes, drug-seeking behavior may be extinguished due to the "cue–no drug" association developing alongside the original "cue–drug" association.

The inhibitory learning that occurs during extinction requires similar encoding, consolidation, and retrieval processes as during the excitatory learning associated with initial acquisition. There are important similarities in the systems and molecular steps that are involved in initial memory formation and extinction (e.g., Ref. 17), but there also are critical differences. For example, initial memory formation and extinction may recruit specific and distinct subregions of the medial prefrontal cortex, amygdala, and NAc (e.g., Refs. 9,18–20). It is thought that these excitatory and inhibitory circuits are usurped during the transition from drug use to eventual addiction (e.g., Ref. 21) and modulate inhibition of drug-seeking behavior during extinction or abstinence. Although extinction treatment (a model of clinical exposure therapy[22]) diminishes drug-seeking behavior, relapse often occurs over time (spontaneous recovery; e.g., Ref. 23), with the presentation of drug-associated cues (reinstatement; e.g., Ref. 24) or after leaving the extinction context (i.e., renewal; e.g., Refs. 25,26). The mechanisms of extinction and relapse are of particular interest in this chapter as they may apply to the treatment and potential prevention of many disorders where extinction may be impaired, including posttraumatic stress disorder (PTSD) and addiction (see Ref. 27).

4. HISTONE-MEDIATED EPIGENETIC MECHANISMS

We now know a great deal about the learning and memory processes that are involved in the establishment and maintenance of addiction. There are several ways in which histone-mediated changes may impact these learning and memory processes. Much of the work on histone modifications in addiction has focused on histone acetylation, but before reviewing those findings, it is important to consider other ways in which histones can be modulated by different molecular events. The study of histone modifications in addiction is in relative infancy compared with the study of these modifications (and epigenetic events in general) in other biological processes, such as cancer.

While epigenetic modifications to the genome are known to change molecular, cellular, and systemic function, they are being discovered as underlying mechanisms to many complex diseases, including developmental, neurodegenerative, and psychiatric (see Refs. 28,29). Much of what is known about epigenetic mechanisms involved in cell biology comes from studies that have focused on the relation between these mechanisms and the causes of and cures for cancer.[30] These basic studies have led to the examination of epigenetic mechanisms in other processes, such as learning, memory, and addiction. The similarities between neural circuits, substrates, and many of the epigenetic factors that create long-term memories and those that cause long-term addiction suggest that common mechanisms are involved (e.g., 31,32). A major focus of current research is investigating how these mechanisms may contribute to developing persistent cellular and molecular changes that may translate into persistent behavioral changes, including lasting suppression of drug seeking.

There are at least five types of chemical modifications (i.e., methylation, phosphorylation, acetylation, poly(ADP-ribosylation), and SUMOylation) that take place on the amino acids of histone tails. These modifications allow access to certain genomic regions to be increased or decreased, which is associated with activation or repression of transcription of specific genes. The enzymes that complete these chemical modifications to DNA, TFs, and histones are typically recruited during development or after some type of stimulation to an organism (e.g., Refs. 3,33). With activity-dependent depolarization of neurons, activation of inter- and intracellular pathways leads to an interaction between these enzymatic coactivators and their substrate (e.g., DNA, TFs, or histones), largely regulating gene and protein expression, ultimately altering system function. The steps in this process are also modified by the physiological state of the organism and the type and extent of stimulation applied.

Research pertaining to histone-mediated epigenetic regulation of addiction has focused largely on the induction of factors downstream of histone modifications that are associated with increases or decreases of drug seeking in rodents. Multiple studies have identified a major role for immediate early genes (IEGs; such as *c-Fos, c-Jun,* and *FosB*), TFs and coactivators (such as cAMP response element-binding protein (CREB) and CREB-binding protein (CBP)), kinases (such as protein kinases A and C) and other proteins (such as Ras) in the plasticity induced during learning or drug use.[2,34,35] The involvement of these key regulatory factors and the related gene targets relies on complex chemical modifications made to both DNA and histone proteins. In the last two decades, increasing recognition for the necessity of these modifications has led to advancements in the field. While both DNA and histone modifications have been shown to interact and rely on each other,[36] each substrate (DNA or histone protein) modification leads to a unique sequence of events.

Multiple types of chemical modifications can be made to histones, each of which is thought to create a novel surface to be recognized by effector proteins and specific downstream events. Although there are many types of modifications made to histones, such as phosphorylation, SUMOylation, ubiquitination, and poly(ADP-ribosylation), the majority of addiction-related research has investigated the effects of altering methylation and acetylation levels on histones 3 and 4.[5] Broadly, research suggests that drug use stimulates many chemical modifications that are needed to activate or repress the transcription and translation of DNA into functional proteins.

Some of these marks can even have opposing effects on transcription. For example, methylation and phosphorylation are thought to participate in both closing and opening of chromatin and potentially mediating the repression and activation of transcription.[4] Alterations to transcription after histone methylation or phosphorylation are based on many variables (e.g., the organism's developmental stage, the onset and duration of stimulation, and the type of tissue, cell, or histone residue that is being targeted for modification[4,37,38]). In contrast, histone acetylation is primarily associated with activated gene transcription with very few exceptions (e.g., Ref. 39). The combination of these marks is in large part thought to be how environmental effects lead to individual variability and how such great diversity can be created through epigenetic regulation. Yet it is still unknown whether histone-mediated epigenetics can be used in the clinic to prevent and treat addiction. Here, we will review the research and remaining questions pertaining to histone-mediated epigenetics in addiction.

4.1. Repressive histone modifications

4.1.1 Methylation and phosphorylation

A repressed state is by and large the default structure of chromatin, preventing abnormal changes to the DNA's code (e.g., segregation, recombination, and replication[40]). The most common way that the tails of each histone remain tightly bound within the chromatin structure, repressing future gene transcription, is through methylation. Methylation is the act of adding or subtracting a methyl group to a substrate or, in the case of histone-mediated epigenetics, to an amino acid in a histone tail. Due to the chemical properties of histones and DNA, epigenetic modification by methylation is only possible on lysine and arginine amino acids and only on amino acids that are bound around or within the tail region of histones 3 and 4. There are three possible degrees of methylation (mono-, di-, and trimethylation), providing further regulation and ultimately determining the outcome on transcription.[41] In addition to the state of methylation, the high degree of specificity to function arises from the residue modified and from the type of enzyme that catalyzes lysine and arginine methylation or demethylation (histone methyltransferases and histone demethylases (HDMTs)). The enzyme complexes created largely regulate histone structure and accessibility to promoter or coding regions on individual genes within DNA. For example, repressive-type di- or trimethylation often occurs on histone 3 (H3) at lysine 9 (K9) or 27 (K27) residue, whereas activation is often associated with trimethylation that occurs on H3K4 or H3K36.[37,42–45] Evidence like this clearly demonstrates how repressive methylation can be used to repress transcription in a complex way. As an example, monomethylation on histone 4 at lysine 20 (i.e., H4K20me1) has been shown to functionally change chromosome condensation and transcription[46] and trimethylation at lysines 9 and 27 on histone 3. To be expected, each of these methylation marks similarly binds nucleosomes together into a more stable and condensed structure and primarily represses overall gene transcription but is associated with different downstream events accomplishing different results through slightly separate avenues. For example, monomethylation of H3K27 and H3K9 may serve to recruit additional silencing machinery and be a substrate for trimethylation that further enhances the repression of noncoding regions of the genome.[47,48]

In addition to the location and type of methylation mark, the timing of histone marks and the host's current modification state may affect the binding of additional regulatory proteins, such as heterochromatin protein 1α (HP1α). HP1α interacts with the methylated tail of lysine 9 on H3 (i.e., H3K9me) and dimerizes with other regulatory proteins. The methylation and additional cross-linking between proteins reinforce the stable and condensed chromatin structure, called heterochromatin. This tightly bound heterochromatin structure increases gene silencing overall. In addition,

HP1α can also be modified, through phosphorylation. Once serine residues on HP1α are phosphorylated, the structure will greatly increase its binding efficiency to H3K9me.[49] In this way and many others, methylation and its associated partners (e.g., HP1α) can result in transcriptional inhibition and decreased genome activity. Similar to phosphorylation of HP1α, histones can also be phosphorylated and increase chromatin condensation, potentially decreasing replication and transcription.[50–53] The details related to repressive effects of phosphorylation on histone tails are not yet characterized well in the addiction field.

4.2. Active histone modifications

4.2.1 Methylation

In contrast to the common repression of gene transcription by histone methylation, a few studies have demonstrated that methylation can in fact activate gene transcription. Trimethylation at lysine 4 on H3[41] and dimethylation at lysines 36 and 79[54,55] contribute to transcriptional elongation and euchromatin, the accessible chromatin structure created to enhance gene activation, transcription, and behavioral changes. Modifications like these three activating methylation marks require different catalyzing enzymes (e.g., methyltransferases Set 7/9, Set2, or Dot1, respectively) and binding proteins.[55] Each enzyme and subsequent mark recruits different downstream effectors and leads to different outcomes on transcription, further demonstrating how histone-mediated regulation is both specific and elaborate. As an example, the methyltransferase and demethylase responsible for di- and trimethylation of H3K4 (i.e., writer MLL1 and eraser kdm5c) are thought to upregulate transcription of the oxytocin receptor and Fos protein in the NAc, mediating methamphetamine-associated memory development and expression.[56]

4.2.2 Acetylation

Histone acetylation occurs on the nitrogen-containing side chain of lysine amino acids and causes the positive charge of histones to detach from negatively charged DNA in the chromatin structure.[57,58] Acetylation levels are largely coordinated through histone acetyltransferase (HAT) and histone deacetylase (HDAC) enzymes. HDACs are the enzymes that remove acetyl groups from amino acid tails of histones. HDACs can be classified into four classes differing by the types of tissues they reside in, where they are active within the cell, the number and homology of catalytic sites, and what substrates and binding partners they interact with to determine functional outcomes.[59] For example, class I HDACs reside within the nucleus of cells, are dispersed ubiquitously throughout the body, have one catalytic site, and have activity on DNA binding TFs and nuclear receptors, signaling mediators, and chromatin-remodeling substrates. Each class and individual HDAC is thought to serve different functions. By decreasing class I HDAC function, alterations in cell survival and proliferation occur, whereas knockout analysis of class II HDACs may localize effects to specific tissue types.[59] Within each class, individual HDACs contribute to a wide range of independent roles, from cardiac function and chondrocyte differentiation to changes in global histone acetylation and gene expression.[59] Additionally, the phosphorylation state of HDACs themselves can determine their permissibility to histones as well. For example, HDAC5 is known to be phosphorylated through activity-dependent mechanisms and after cocaine administration and then exported out of the nucleus, decreasing activity at specific histone sites.[60] Correspondingly, the repression of genes, such as *NR4A1*, a nerve growth factor involved in inflammation and cell survival, is increased by dephosphorylation of HDAC7. It is still unclear how HDAC function is selectively recruited and how sequence specificity of histone tails helps determine and coordinate these regulatory factors.

4.2.3 Phosphorylation, ubiquitination, SUMOylation, and poly(ADP-ribosylation)

These posttranslational modifications are well understood in other areas of research, but less studied with respect to addiction physiology. Phosphorylation of serine 10 on histone 3 has been shown to induce acetylation at a nearby lysine site (H3K9) and potentially others (e.g., H3K14) and subsequent transcription.[4,61,62] It is here that histone phosphoacetylation is thought to occur, subserving a "subset of rapid transcriptional responses" for gene induction.[62] These are thought to be mediated by mitogen-activated protein kinase and often lead to CREB and CBP/p300 activation. Although little research has connected histone phosphorylation or phosphoacetylation to addiction (to be discussed in Sections 4.2 and 4.3), cancer research recently reported evidence that androgen stimulation can result in histone phosphorylation on histone 3 at threonine 11 (H3T11), recruitment of H3K4me3 and H4K16ac, and subsequent activation of androgen receptor target genes and proliferation of prostate cancer.[63] Similar to these studies, these phosphorylative steps may be initiating or potentiating other posttranslational cascades that are recognized to occur during or after drug use (i.e., acetylation and methylation of histones).

Ubiquitination is a key regulator for several histone modifications. Histone ubiquitination is the addition of a covalently bonded ubiquitin protein to the lysine residue on the N-terminal tails of histones. The ligation of one or more ubiquitin subunits onto lysine residues and the target proteins they reside in will be marked for further protein trafficking or degradation.[64,65] The addition of these bulky ubiquitin moieties to substrates, such as histones, initiates degradation or activity, communication, or location changes. The main catalyst recruited for the ubiquitination system is Rad6, the enzyme dedicated to ubiquitin-like conjugation (e.g., the addition of ubiquitin) of substrates.[44] Ubiquitination often works in a phosphorylation-dependent manner to control gene transcription. It may also be a signal for active versus inactive chromatin and recruit methyl and acetyltransferases to assist in silencing or activating gene transcription at specific loci, respectively. In addition to catalyzing ubiquitination, the Rad6 enzyme likely plays a role in recognizing active versus inactive substrate sites on chromatin.[44]

SUMOylation is similar to ubiquitination, as it results in the addition of bulky peptides to substrates such as histones. SUMO stands for small ubiquitin-like modifier, one that can recruit HDACs and HP1 and lead to potent transcriptional repression.[55] This modification not only has been identified on a few histone sites so far, shown to block activating events like acetylation or ubiquitination, but also occurs on TFs, indirectly repressing histone acetylation and gene transcription.[55,66] In addition, the phosphorylation state of substrates (e.g., various kinases, TFs, and histones) can inhibit or enhance SUMOylation.[66] While the temporal and spatial details are limited to date, it is clear that complex interactions exist between SUMOylation and other histone marks (e.g., phosphorylation, acetylation, and ubiquitination) to repress transcription.

Histone poly(ADP-ribosylation) is the addition of one or more ADP (adenosine diphosphate)-ribose moieties to acceptor sites such as lysine, arginine, glutamic acid, and aspartic acid of histones. Nuclear proteins, such as histones, interact with poly(ADP-ribose) polymerase 1 (PARP1) to get mono or poly-ADP-ribosylated. The activation of PARP1 can mediate chromatin structure, gene transcription, and environmental stimuli responses of cells, during both development and adulthood.[67] Accordingly, it is well established that PARP1 is required for long-term memory formation by targeting (e.g., p53 and Fos) and binding to a variety of TFs (e.g., NF-κB and AP-2), each involved in plasticity or learning and memory processes.[68,69]

4.3. Combinatorial modifications

As noted briefly above, these modifications may act alone or in concert with other modifications to create a complex code that determines individual gene regulation. For example, H3K9me3 initiates and maintains repression of transcription by antagonizing active modifications (e.g., H3K9ac, H3S10p, and H3K4me3[54]). In addition, extensive coordination occurs between histones and DNA to modify the epigenome, facilitating, impairing, or neutralizing transcription. By depleting the enzyme that poly-ADP-ribosylates proteins (PARP1), H4K20me3 will decrease, and further deplete ubiquitin ligase UHRF1 and DNMT1 (DNA methyltransferase 1) expression.[70] Without the maintenance of this DNA methylation, the nucleosome assumes a euchromatic structure and allows abnormal transcription. Another example of DNA- and histone-mediated interactions was demonstrated in fetal rat cortical neurons.[71] Stimulation and calcium influx of these neurons result in phosphorylation of methyl CpG (cytosine–phosphate–guanine)-binding protein 2 (MeCP2), releasing it from methylated CpG sites of DNA and ultimately decreasing methylation and increasing acetylation at lysine 9. Early results from Jones *et al.*[139] also demonstrated that MeCP2 binds methylated DNA and recruits HDACs to further stabilize transcriptional repression of chromatin. Therefore, when MeCP2 is bound to DNA, it is responsible for global gene repression, and when released, HATs and demethylases help remove methylation and acetylate nearby sites (i.e., H3K9) to increase activity at specific gene promoters and increase protein expression (e.g., promoter region of brain-derived neurotrophic factor (BDNF)[71]). While this is just one example of how DNA and histone-mediated epigenetics can regulate chromatin accessibility and relevant learning and memory proteins, many examples of this coordination exist (discussed further in Ref. 33).

5. HISTONE-MEDIATED ACTIONS IN DRUG ADDICTION

With the use of animal models and advancing technology, researchers are trying to uncover the complex environmental and genetic mechanisms that underlie addiction. Although significant differences exist between human and animal conditions of drug craving, seeking, and taking,[72] the research field is focused on delineating each aspect of addiction, in the hopes of piecing together a pathway for improved treatment and prevention.

Because drugs of abuse often change cellular and systemic activities in animals and humans, simple measures of cellular and locomotor sensitization or tolerance are used to measure changes in physiology (e.g., receptor function) and overall behavior after acute or chronic drug administration. In addition, animal models like conditioned place preference (CPP) and operant self-administration are commonly used to investigate the reinforcing effects of drugs and infer the strength of drug–cue or drug–response associations. Similar to humans, animals will associate the reinforcing effects of drugs with stimuli (i.e., places) and behaviors (i.e., voluntary responses that lead to drug delivery) that are paired with a drug experience. Subsequently, if reintroduced to drug-paired stimuli or the opportunity to obtain drug reinforcement, animals and humans alike will retrieve these memories and demonstrate drug craving, seeking, or taking behavior.

The various histone modifications discussed above (i.e., Section 4) contribute to many diseases, including drug addiction, and are initiated and terminated for many reasons.[33,73,74] Similar to other diseases, histone-mediated changes can influence and be influenced by a patient's or subject's genetic background, surrounding environment, and stage of development or disease. Within the drug abuse field, epigenetic researchers have predominantly discovered changes that occur to histone methylation and acetylation status (e.g., Ref. 75).

5.1. Methylation and addiction

One of the more prominent and recent discoveries on the mechanisms underlying addiction described how the epigenetic response to an initial or habitual dose of drug administration is often different (see Fig. 3.1). After just one administration of cocaine (acute exposure shown in Fig. 3.1), the methyltransferase responsible for methylating H3K9 sites, G9a, is increased. Increased methylation at this site results in greater binding of G9a to the IEG *FosB*, an effect that seems to be counteracted after repeated cocaine use, where G9a levels and *FosB* binding decrease.[76] As noted previously, methylation often leads to a heterochromatin, or an inaccessible structure, increasing the likelihood of decreased transcription. While it is known that ΔFosB, a product of the *FosB* gene, accumulates with repeated cocaine exposure and is associated with increased cocaine reward,[77] removing this G9a hindrance at *FosB* sites enables ΔFosB to be increased, perpetuating the accumulation of ΔFosB and the addiction cycle.[76]

Additional studies have demonstrated that chronic exposure to drugs of abuse, such as cocaine and opioids, reduces dimethylation of H3K9 by

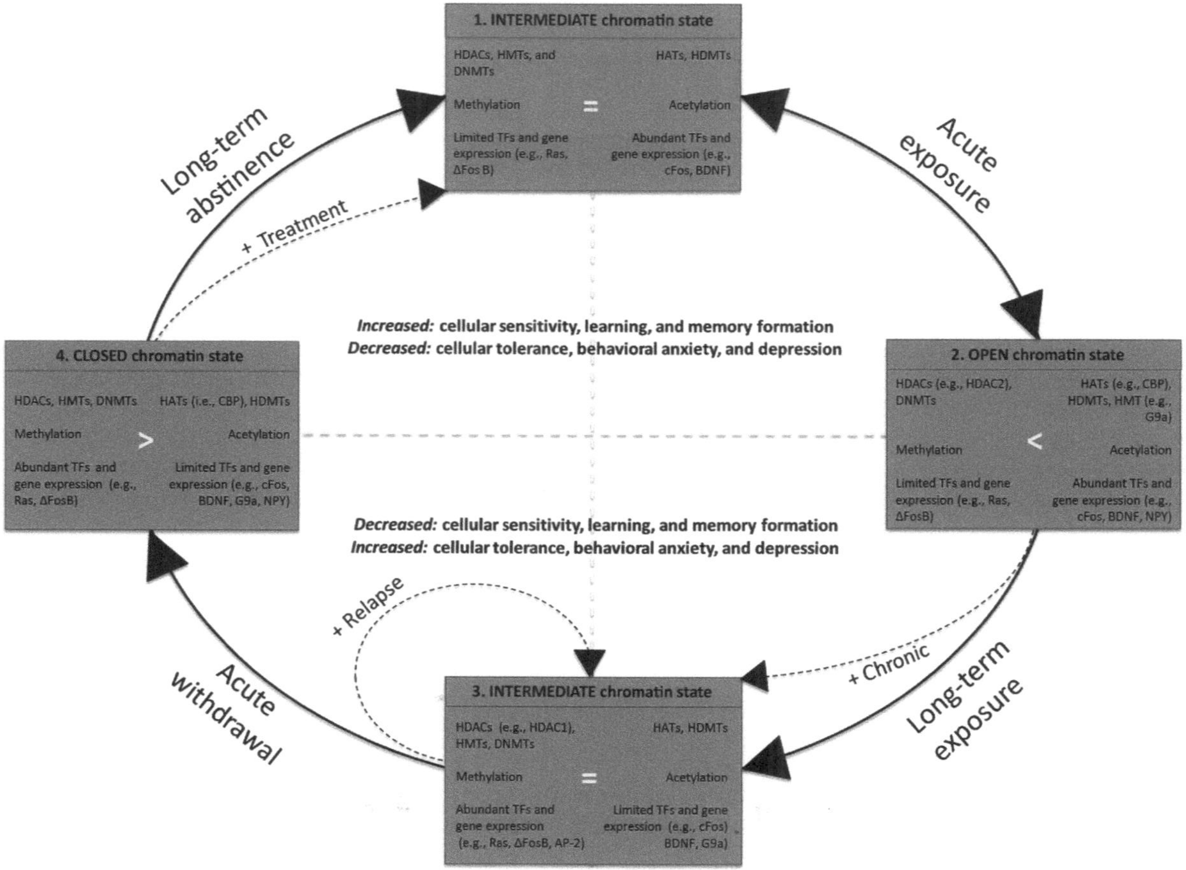

Figure 3.1 Epigenetic changes in the cycle of addiction. Some of the potential chromatin dynamics are shown for the cycle of addiction, which moves from acute exposure, to chronic drug taking, to withdrawal, to abstinence and recovery, back to acute exposure in cases of relapse, which begins the cycle again. Three potential states of chromatin (i.e., OPEN, CLOSED, or INTERMEDIATE) and their associated nuclear changes are depicted within four small gray boxes (numbered 1–4). These chromatin states create a more accessible (Box 2), inaccessible

(Continued)

Figure 3.1—Cont'd (Box 3), or intermediate (Boxes 1 and 3) structure for DNA to be accessed and transcription to take place. The top half of this figure signifies chromatin in a more "adaptive and responsive" state. The bottom half signifies chromatin in a more "inflexible and unresponsive" state (splitting states 2 and 4 into both of these categories evenly). The left half of this figure signifies a more "heterochromatin" state, while the right half signifies a more "euchromatin" state (splitting states 1 and 3 into both of these categories). Dashed lines (e.g., +chronic, +relapse, and +treatment) represent the potential for associated changes to be accelerated by rate and/or intensity. Box 1 (INTERMEDIATE chromatin before drug intake) represents a basal chromatin state with normal transcription (determined primarily by genetic and previous environmental interactions). In this state, the chromatin and associated nuclear changes are well balanced and highly regulated. *Box 1 → Box 2 transition*: With acute exposure to stress or drugs of abuse, brief and reversible changes (see bidirectional arrows) occur to select histone and DNA regions (increased histone acetyltransferases like CREB-binding protein [CBP], acetylation, DNA accessibility, and learning and memory-related gene transcription). Box 2 (OPEN chromatin) represents the change that occurs with a single or acute insult to the system (e.g., acute stress or drug exposure). Chromatin expands, releasing repressive marks and tipping the balance of epigenetic regulation toward those associated with gene activation. *Box 2 → Box 3 transition*: With repeated exposure to stress or drugs, a prolonged and less reversible change occurs to select histone and DNA regions. Box 3 (INTERMEDIATE chromatin after drug intake) represents chromatin with dysregulated histone enzymes, marks, TFs, and transcription. In this state, similar process occurs as in Box 1, yet the location where histone modifications occur, the type of modification, and the effect that histone modifications have on cellular and behavioral outcomes is altered to positively reinforce this chromatin state. The balance of regulation is shunted away from promoter regions that are associated with learning and adaption (cFos and BDNF) and shifted toward promoter regions that are associated with this altered chromatin state and positively reinforce this altered gene regulation (e.g., Ras and ΔFosB). These changes are thought to induce increased cellular tolerance and maladaptive behavior. *Box 3 → Box 4 transition*: With acute drug abstinence, brief and reversible changes occur to select histone and DNA regions (increased histone deacetylases, methylation, DNA inaccessibility, and decreased gene transcription) in an attempt to rebalance the previous dysregulation. Yet after chronic or repeated insults to the system, recent drug abstinence induces withdrawal-associated effects (e.g., anxiety and depression), making the organism increasingly susceptible to relapse rather than recovery and long-term treatment. Box 4 (CLOSED chromatin) represents the change that occurs with acute abstinence (without relapse) and the associated withdrawal from drugs of abuse. Here, chromatin begins the process of rebalancing enzyme levels, histone marks, and gene transcription by generally increasing the repression of prior imbalances related to addiction. The previous epigenetic and behavioral changes placed on the system (e.g., positive feedback of ΔFosB and behavioral depression) make this process slow, the system resistant to rebalancing and deprived of necessary proteins to counteract this state. As chromatin becomes more condensed, regulation is increased (although exceptions to this mechanism exist, such as decreases to repressive methylation with withdrawal; noted in Section 5.3). *Transition from Box 4 → Box 1*: With repeated and long-term abstinence from drug use, a prolonged and less reversible change occurs to select chromatin regions, rebalancing the location, type, and effect that histone modifications have on cellular and behavioral outcomes, recovering to a more normal and highly regulated level of transcription. (See the color plate.)

decreasing G9a and GLP (G9a-like protein) enzymes in the NAc of mice.[56,75,78] Similar effects occur in the mouse cortex and in cultures of human lymphocytes after repeated nicotine treatment.[79] In parellel, G9 levels decrease as drug taking increases with chronic treatment (i.e., as administered and measured with an animal model of operant drug self-administration), but in contrast, are increased after acute drug administration[76,78] potentially leading to initial responses in drug-seeking and drug-sensitization behavior (as measured by animal models in CPP and locomotor sensitization paradigms). This suggests that increased levels of dimethylation of H3K9 impair drug reward initially but are mitigated by repeated drug use.

Trimethylation of H3K9 has also been shown to play an important role in the addiction process. While H3K9me1 is known to reside in the euchromatin and be correlated to gene activation, H3K9me3 occurs in nongenomic or heterochromatin regions of the DNA and is correlated with gene repression.[37,140] Repeated cocaine treatment increases the expression of this specific type of methylation, resulting in enhanced expression of transposons and typically silenced (heterochromatic) regions of the DNA by decreasing repressive methylation within the NAc of mice.[47] This work and others noted above suggest that repeated treatment with drugs of abuse causes the derepression of previously silenced DNA regions by inhibiting methylation at key nongenomic and genomic sites. This is likely a key factor contributing to the altered gene expression and impaired physiological function after long-term use of drugs and with addiction. Importantly, these alterations are thought to release the brakes placed on transcription and lead to a more permissive epigenetic environment.[80]

Covington *et al.*[81] further investigated if dimethylation on H3K9 could be related to the depressive-like phenotype expressed after chronic cocaine administration and social defeat stress (a paradigm inducing anxious, stressful, and depressive characteristics in animals[82]). As expected, their results suggested that chronic cocaine administration led to increased vulnerability to the detrimental effects of social stressors and that this was due to the removal of G9a, GLP, and subsequent demethylation of H3K9me2 and enhanced Ras–CREB signaling. This is likely a key mechanism for chromatin to be opened and the expression of downstream proteins associated with addictive behavior to increase (e.g., Ras G-proteins and ΔFosB TFs). In contrast, increased BDNF in rats leads to greater drug reinforcement[83] and is decreased with chronic cocaine use in humans.[84] BDNF's downregulation in reward-related regions of the brain (e.g., VTA and NAc) after chronic cocaine use and its replenishment during abstinence[84] are likely compensatory mechanisms of cocaine's effects. Interestingly, if the regulation of this factor contributes to the rewarding properties of cocaine and is increased during abstinence, this may account for slower rates of CPP extinction reported in rats[83] and the positive correlations between BDNF, abstinence, anxiety, and depression during early abstinence.[84] These types of changes in gene regulation are often enabled by the combination of decreased methylation status and subsequent increased histone acetylation.[85] Coordinated mechanisms between histone methylation and acetylation will likely be targeted for treatments of addiction and other disorders in the future.[86,87]

5.2. Acetylation and addiction

As more diseases are being attributed to novel histone-mediated mechanisms (e.g., palmitoylation, isomerization, and deimination[42,88–90]), alterations involving histone acetylation have been leading the charge in addiction research. Major advances in understanding how occasional drug use can manifest into chronic problems have largely been due to the use of pharmacological agents that target HDACs. Initial investigations with HDAC inhibitors focused on anticancer activity,[91–93] but more recent work has examined the potential of HDAC inhibitors in treating a variety of psychiatric disorders (i.e., addiction, PTSD, and depression; reviewed in Ref. 75). One of many examples of this is with trichostatin A (TSA), a general HDAC inhibitor with antitumor activity,[94] that has been shown to decrease the motivation for and intake of addictive drugs.[95] Such a finding has been observed with other HDAC inhibitors and variations to the time and route of drug administration.[63,96,97]

In an operant measure of drug-seeking behavior, Romieu *et al.*[95] demonstrated that rats that voluntarily self-administered high levels of cocaine would decrease responding when administered with this nonspecific HDAC inhibitor, TSA. In addition, repeated administration of TSA was shown to stably and selectively decrease cocaine administration and not sucrose intake while reducing cocaine-induced locomotor sensitization. This was mediated through decreases in HDAC deacetylation activity in the prefrontal cortex and NAc, either with or without cocaine administration. These results and others[98,99] led to the potential use of HDAC inhibitors in relapsing addicts.[95] Although the circuits mediating these effects remain to be fully described, a focus is on the NAc, where drugs of abuse increase acetylation, locomotion, reinforcement, and reward.[100–104]

Work from these groups and others has also demonstrated that chronic drug use (e.g., with cocaine or alcohol abuse) will initiate decreases in HDAC activity in reward- and learning-related brain regions (e.g., NAc, prefrontal cortex and hippocampus[95,105]), creating a more permissive genome for drug regulation to be imparted. Renthal *et al.*[106] discovered a potential mediator for the transition from dysfunction or drug abuse to disease and drug addiction with HDAC5. HDAC5 is a class II HDAC that is phosphorylated minutes after activity-dependent stimulation (e.g., initial drug use and stress) and is exported from the nucleus, allowing gene transcription and behavioral adaption to take place. Without additional insult, HDAC5 will be replaced in the nucleus within 24 h, capping gene transcription. Chronic stimulation, such as chronic cocaine administration or chronic social defeat stress, results in long-term HDAC5 inhibition, increases in cocaine- or stress-associated target genes and proteins (e.g., Ras and ΔFosB), and dysregulated sensitivity to subsequent challenges. These data demonstrate the importance of balancing acetylation to maintain flexible behavior.

As occasional drug use turns chronic, many cellular and behavioral changes occur (long-term exposure shown in Fig. 3.1). In light of the role that HDAC5 plays in chronic but not acute drug treatment or stress, the TF ΔFosB is also thought to mark the transition from abuse to addiction.[77] With chronic amphetamine treatment, ΔFosb is increased, leading to the recruitment of HDAC1 and attenuation of the IEG, *c-Fos*.[77] In concert, dimethylation of histone 3 at lysine 9 is increased (recall that this type and location of methylation is repressive in nature). These data further demonstrate that chronic drugs (i.e., amphetamine) not only swap methylation for acetylation at the *c-Fos* promoter within the striatum (a region highly important in addictive behavior) but also lead to decreased expression of a methyltransferase (KMT1A) necessary to remedy this imbalance. As ΔFosB increases with repeated drug use, G9a (another methyltransferase) is built to decrease the levels of this addiction marker by binding to the *FosB* promoter and suppressing its expression.[76] As one can see though, the coordinated tools put in place to rectify such epigenetic imbalances are in competition with drug effects that are inherently self-perpetuating.

While drug use impairs HDAC activity and positive feedback of ΔFosB (by ΔFosB) is initiated during chronic drug use (reviewed in Ref. 107), additional changes compound these imbalances. For example, an acute administration of alcohol or cocaine will briefly increase H4 acetylation and H3 phosphoacetylation (previously associated with active IEGs, *c-Fos* and *c-Jun*[62]). Kumar *et al.*[100] demonstrated mechanistic differences between acute and chronic covalent chemical changes in the brain. They determined that histone acetylation and phosphoacetylation influenced gene regulation in the striatum, an addiction-mediating region of the brain, and behavior in mice and rats after acute and chronic cocaine administration (by investigator-initiated intraperitoneal injections and subject-initiated self-administration infusions). Cocaine induced acetylation (on H4) and phosphoacetylation (on H3) at specific gene promoters, with acute effects at the *c-Fos* and *FosB* sites on H4 and chronic effects at the *FosB*, *cyclin-dependent kinase 5 (Cdk5)*, and *BDNF* sites on H3. This creates an overall pattern of hypoacetylation and desensitization on H4 at the *c-Fos* promoter but an exaggerated state of acetylation on H3 at *FosB* promoters (only partially desensitizing ΔFosB[77,100]). Similar patterns of gene regulation have been demonstrated with *c-Fos* and *BDNF* in the hippocampus (rather than *c-Fos* and *FosB* in the striatum) after acute and chronic electroconvulsive seizure induction (a procedure shown to decrease long-term plasticity in the hippocampus of rodents and be an effective treatment for depression in humans[108]).

Similar to the previous studies mentioned, alcohol can also modify acetylation patterns with temporal and spatial specificity.[109,110] Acute doses of alcohol can alleviate minor levels of stress or anxiety-like behavior,[111] possibly because of alcohol's ability to decrease HDAC levels in the amygdala (a brain region necessary for affective and emotional associations[112–114]). Pandey *et al.*[113] demonstrated that decreases in HDAC activity (i.e., following acute ethanol) correspond to increases in H3 and H4 acetylation and increases in anxiolytic and plasticity-related protein levels (i.e., CBP and neuropeptide Y (NPY)). In contrast, increases in HDAC activity and subsequent decreases or rebalancing of H3 acetylation, CBP, and NPY likely mediate alcohol withdrawal and associated anxiety.[113] As previous data would predict, inhibition of HDAC activity (by TSA) normalizes the sharp decreases in H3ac, H4ac, and NPY expression and the corresponding anxiogenic effects of withdrawal.[113]

Although there are nuances, the majority of research thus far supports the idea that drugs of abuse and various stressors generally repress protective methylation and enhance permissive acetylation, which can be detrimental to adaptive behavior and lead to addiction and depressive-like behaviors. Interestingly, these two common mechanisms can lead to an increase in transcription, yet the timing and location of these modifications play a critical role in their behavioral outcomes.

5.3. Combinatorial modifications and addiction

While histone modifications are merely a part of the physiological changes that occur with drug addiction, it is becoming accepted that each histone modification plays a role in the development, maintenance, or potential treatment of drug addiction. Therefore, understanding the upstream and downstream coordinations of each modification is becoming increasingly necessary and elaborate. As one might expect, dopamine receptor agonists elicit rewarding effects, but the combination of dopamine agonists and HDAC inhibitors seems to compound these effects, leading to a synergistic increase in cocaine-induced locomotion or sensitization and CPP.[102] Research by Schroeder *et al.*[102] demonstrated that a class I/II HDAC inhibitor (sodium butyrate) and SKF82958 (D1 receptor agonist) increased H3 phosphoacetylation (i.e., phosphorylation at serine 10 and acetylation at lysine 14) in striatal homogenates and increased deacetylation in the substantia nigra/VTA at the promoter regions of *BDNF* and tyrosine hydroxylase (the rate-limiting enzyme for synthesis of catecholamines like dopamine and norepinephrine). In another study, a dopamine D2 receptor antagonist was found to induce similar changes.[115] Importantly, the molecular effects in the Schroeder *et al.*[102] study were sensitive to acute versus repeated administration of these drugs and CPP effects only emerged when the drugs were administered directly after cocaine-induced place preference (presumably during consolidation of CPP learning). This highlights two important points: (1) the transient molecular effects of epigenetic modifications that are known to change after repeated drug use and (2) the experience or activity-dependent nature of histone modifications.

Research by Qiang *et al.*[116] demonstrates how methylation (i.e., repressive di- and tristates) and acetylation coordinate to compound the effects of histone modifications. Neuron cultures of mice had large increases in activating acetylation at H3K9 that coincided with large decreases in repressive methylation during the time of withdrawal from ethanol. Surprisingly, global and local downregulation of HDMTs (i.e., G9a) at the NMDA receptor gene (*NR2B*) rather than changes in global or local HATs or HDACs seemed to underlie this effect, underscoring the idea that these modifications work in concert to regulate or dysregulate the system at specific chromatin sites. Correspondingly, Sheng *et al.*[117] demonstrated that intracranial administration of an NMDA antagonist (MK-801) into the NAc (but not the medial prefrontal cortex) decreased H3 phosphoacetylation and acquisition of heroin CPP, while heroin CPP dose dependently increased H3 phosphoacetylation in the NAc (but not the medial prefrontal cortex). TSA (HDAC inhibitor) infusions prior to each CPP acquisition session facilitated heroin's effects, further increasing H3 phosphoacetylation in the NAc and CPP. Research like this clarifies ideas that certain brain and chromatin regions and types of histone modifications are targeted based on drug type and administration patterns.

As we better understand how cocaine and alcohol affect methylation and acetylation, appreciation grows for modifications involved with other drugs of abuse. For example, hints are emerging that poly(ADP-ribosylation) and ubiquitination may also contribute to the cellular changes that occur with drug abuse. Cigarette smoke extract leads to an overall increase in H4 acetylation, due to ubiquitination and subsequent degradation of HDAC2.[118] In addition, the dopamine transporter (DAT), implicated in cognition, affect, behavioral reinforcement, and motor control, can be tagged by ubiquitination[119] and lead to altered endocytosis and degradation levels. Although it is known that drugs of abuse alter the function of DATs, the full mechanism and histone modifications likely behind these alterations are clearly still being discovered. In addition, it is established that the primary polymerase instigating poly(ADP-ribosylation) polymerase (PARP) is a coactivator of AP-2-mediated transcriptional activation,[120] but no direct link has been formed between histone ribosylation and transcriptional changes by drugs of abuse thus far. However, the overexpression of AP-2 recruits the coactivator PARP and recent evidence attests to the idea that acute and repeated morphine exposure locally increases AP-2 in the neurons of mice hippocampi (i.e., specifically at the postsynaptic density). This is thought to contribute to the dysregulation of hippocampal plasticity, by specifically targeting glutamatergic synapses (those with AMPA and NMDA receptors). Although many factors contribute to the synaptic changes noted in this and other studies, PARP and increased histone poly(ADP-ribosylation) may be other mechanisms of drug-induced changes usurping learning and memory functions. While many of the links between ubiquitination, ribosylation, histones, and addiction have yet to be uncovered, these marks (like others) present plausible mechanisms by which drugs of abuse, epigenetic processes, and long-term changes take place.

5.4. Summary of histone-mediated epigenetic regulation in addiction

Fundamental theories about addiction evolved from the epigenetic research summarized here and elsewhere.[31,32,111,121] This system of regulation enables an infinite level of control and flexibility on gene regulation. The interaction between specific sequences of DNA, specific histone proteins, and other regulatory factors induced by the environment leads to highly controlled genomic outcomes, including increased or decreased transcription of selected genes. While initial drug use is now thought to prompt the brief opening and closing of chromatin and adaptive behavior, repeated drug use seems to decrease the system's plasticity as cellular tolerance takes over. This broadens the time and resources needed for each epigenetic modification and narrows the adaptiveness of the system. Without treatment, epigenetic modifications will become perpetually dysregulated and behavioral vulnerability will linger.

6. DEBATES AND CONSIDERATIONS

Because histone-mediated epigenetics in addiction is such a new field, there are many unresolved issues that are currently being debated in the literature. Some of these are general debates that are common in all fields in which epigenetic mechanisms are studied. Other debates are more specific to the addiction and learning and memory fields, where very different theoretical approaches have been offered for defining epigenetic mechanisms, the nature of cellular memory and how it relates to psychological memory, and the nature of psychological processes in general, which are open to many different interpretations. Our goal in this section is to present a broad overview of some of these debates. We cannot provide resolution for most of these issues, but we can survey the literature and provide suggestions about the direction of research in these areas.

6.1. Are histone marks causal or correlational to downstream processes and do histone modifications change the structure and function of the genome?

Histone modifications are a contributing factor in a larger dynamic process that regulates accessibility of DNA and transcription.[122] By changing the charge between nucleosomes and nearby DNA, histone modifications help maintain an unraveled or raveled chromatin structure and docking area for regulatory modules. This account of histone marks highlights their many correlated and few causative roles in transcriptional regulation.[45] As partial justification for a correlative role, histone-modifying enzymes often have effects on other substrates besides histones, leading to multiple effects on the genome, rather than direct effects on transcription only. In addition, it seems that histone modifications play a larger role in stabilizing nucleosome occupancy and position rather than recruiting regulatory factors to DNA. It is unlikely that modifications, like methylation or acetylation, recruit, organize, and direct downstream binding and function of other molecules, since they have minimal binding affinity to most binding modules or regulatory factors.

On the other hand, modifications likely help maintain a particular chromatin state and stability or instability by altering nucleosome dynamics. This presumably helps to ensure high or low accessibility of DNA by other factors to influence transcription. Nucleosomes are thought to impede the binding of elements to DNA. Therefore, by maintaining the occupancy and position of nucleosomes, histone modifications contribute to changes in chromatin landscape, being one of the many events that alter transcription regulation. This is in contrast to the idea, or overgeneralization perhaps, that histone modifications themselves increase and decrease DNA accessibility and are directly activating and repressing transcription (reviewed in Ref. 38). This logic corresponds to the idea that histone modifications are not only creating a simple change of charge between DNA and histones to maintain the overall structure of nucleosomes but also actually creating binding sites for regulatory elements necessary for transcription, with specific location and marks combining to recruit specific regulatory proteins and outcomes.[42,68,123]

6.2. What is the relation between memory at the cellular, organismal, and transgenerational levels?

One of the reasons that epigenetic approaches to addiction are potentially so promising is that successful treatment of addiction requires long-term changes in behavior. Epigenetic changes are associated with long-term changes at the cellular and molecular levels, such as cell fate and cellular memory. But what is the relation between cellular memory and long-term memory that is distributed among different neurobiological circuits? What is the relation between memory at the neurobiological level and memory at the transgenerational level? These are key questions that are not yet resolved.

At the level of the cell, histone-mediated epigenetics may create tags that allow the transcriptional machinery to operate on certain DNA sequences. This creates a specific cellular memory that may result in increased rates and levels of transcription and translation the next time the circuit is activated. This cellular memory need not correspond directly to memory on the level of the whole organism because psychological memories are widely distributed and are controlled by many molecular processes. But the cellular memory that is created by histone marks in various neurobiological circuits triggers the downstream events that are needed to solidify and maintain these organismal-level memory circuits.

Can a histone-mediated epigenetic memory be transferred across generations? This is a critical question that requires an agreed upon definition of epigenetics. This is often disputed, largely on the basis of whether posttranslational changes can be lasting and inherited from cell to daughter cell or parent to offspring.[124] While the term genetics implies the necessity of inheritance, less traditional views are incorporating the idea that changes to the transcription and translation of the genetic code may be inherited along with the DNA code itself. Although some studies have demonstrated that histone-mediated alterations in parents can lead to effects in subsequent offspring during development,[125,126] few addiction-related studies have tested if the offspring can inherit increased risk or protection from parental modifications due to drug use.[73]

In addition, the definitions of trans- and intergenerational inheritance have been broad. As defined by Heard and Martienssen,[126] intergenerational effects are those inherited *in utero* through parental effects or stimulus exposure. This is in contrast to epigenetic effects that are inherited generations later, without exposure to the stimuli that epigenetically altered gene expression or function initially, generations earlier. Few studies to date have demonstrated that drug use alters the information inherited by the offspring and subsequent brain function (e.g., Refs. 127,128) and that this may be mediated through histone modifications.[128] In this last study, a cocaine-resistant phenotype (i.e., slower rate of acquisition and decreased motivation to administer high doses of cocaine as measured by a progressive ratio schedule of reinforcement) was inherited by male, but not female, offspring of parents with extensive and voluntary use of cocaine. The authors showed that this effect was reliant on H3-mediated increases in acetylation levels on the *BDNF* promoter, subsequent increases in BDNF mRNA in the medial prefrontal cortex, and BDNF protein expression. A review by the same group emphasized the likelihood of epigenetic inheritance involved in many diseases, but reiterated that further research on the inheritance of posttranslational modifications and drug abuse was needed.[141]

Evidence for human trans- and intergeneration epigenetic inheritance is still lagging and a key challenge is to demonstrate that experience-dependent changes, such as those outlined in this chapter, can be passed through multiple generations. One study by Norrholm *et al.*[129] demonstrated that PTSD patients with an alleged PTSD risk genotype (i.e., Met/Met single-nucleotide polymorphism) had greater fear to an experimental safety signal and were unable to extinguish this fear after training compared with PTSD patients without the risk genotype. This Met/Met versus Val/Met or Val/Val single-nucleotide polymorphism at the catechol-O-methyltransferase (*COMT*) gene likely contributes to increased forebrain dopamine levels[130,131] in carriers. Accordingly, the Met/Met genotype was associated with greater DNA methylation at CpG sites that were also associated with patients experiencing enhanced fear to safety signals. Another study in rodents demonstrated that fear conditioning in an F0 generation mouse can lead to rapid and specific fear learning in F1 and F2 generation mice.[132] Studies like these (e.g., Refs. 128,129) will contribute to understanding how epigenetic and DNA-based modifications can be inherited and influence health and behavior in subsequent generations.

6.3. What theoretical processes are altered by drugs that target histone modifications during treatment of addiction?

Much of the work reviewed here on histone-mediated epigenetics and addiction points to lasting effects on the levels of the cell and behavior. The most widely demonstrated effects on addiction-related memory processes come from studies showing that administration of HDAC inhibitors can promote memory and synaptic plasticity. When we think of treatments for addiction, however, we are often interested in the ways in which memories associated with drugs can be weakened by some pharmacological treatment.

As noted in Section 3.2, when memories are retrieved, there are multiple theoretical processes that may be triggered. The memory itself may become labile, necessitating a postretrieval reconsolidation process that centers on the original drug-associated memory. However, if the expected drug is not administered, the behavior may begin to show extinction—the absence of the expected drug leads to a weakening of behavior through the development of an inhibitory memory. A consistent finding from the literature on histone-mediated epigenetics is that HDAC inhibitors paired with drug-related memory retrieval will decrease drug-seeking behavior and weaken subsequent relapse. The question at a theoretical level is "why does this occur?" The answer to this question has focused on both extinction and reconsolidation, with the distinction between them often boiling down to assumptions that are made about these theoretical processes.[133]

On one level, behavioral evidence in favor of an impaired reconsolidation account often comes in the form of persistently weakened behavior. Because extinction is often transient, with the response showing spontaneous recovery with time, renewal with context changes, or reestablishment with exposure to drug, any persistent effect is interpreted as an effect on reconsolidation, rather than extinction. Work with HDAC inhibitors, however, challenges this behavioral definition. Because HDAC inhibitors have been shown to promote memory in a variety of behavioral approaches, it is not difficult to imagine how long-term suppression of drug-seeking behavior could result from an enhanced extinction process. Increased histone acetylation and the permissive state of chromatin that it creates should strengthen the consequence of extinction, thereby transforming a potentially weak and transient behavioral experience into one that is long-lasting. Thus, a long-lasting effect on extinction is entirely consistent with an enhanced extinction effect.

On another level, one might examine the molecular processes that are thought to control reconsolidation and extinction. The thought here is that because extinction enhancements involve new memory formation and reconsolidation, impairments involve memory elimination and molecular events associated with memory storage and erasure should be uniquely associated with extinction and reconsolidation, respectively. This is a reasonable way to deal with this issue, but the challenge is that we as a field do not really know which molecular processes are specifically associated with memory storage or erasure. Identical molecular evidence has been offered for enhanced extinction and impaired reconsolidation (see Ref. 134) suggesting that the field does not yet have a handle on how these processes underlie memory.

Perhaps the most promising level of analysis is to examine the circuits that are activated by histone-mediated enhancements in extinction. The line of thinking with this approach is that if circuits that regulate extinction are hyperactivated by a drug such as an HDAC inhibitor, then this could be taken as evidence that extinction has been enhanced. Indeed, research from our laboratory has used this approach to identify how extinction circuits are regulated by HDAC inhibitors.[135] Again, the challenge here is that the circuits that mediate initial consolidation, reconsolidation, and extinction are highly overlapping and interactive. As more is understood about these circuits (e.g., specific amygdala projections that mediate extinction and neural substrates recruited by different contextual and developmental influences), this approach will become even more powerful in distinguishing between enhanced extinction and impaired reconsolidation processes.

Beyond extinction and reconsolidation, future work on histone-mediated changes in addiction-related behaviors will need to identify other psychological processes that are altered by these mechanisms. For example, behavioral research has identified several critical variables in the establishment and maintenance of drug-seeking behavior. One critical variable is the prediction error between the abused substance and the cues in the environment, with unexpected outcomes (or absences of outcomes in the case of extinction) having a large impact on the effects of the drug itself[14] and the strength of the cue–drug association.[136] A second variable is the context in which drug seeking occurs—even after long periods of abstinence, exposure to a drug-associated context is enough to trigger relapse of drug seeking. Although there is a great deal known about how these and other behavioral variables influence drug-seeking behavior, very little is known about how epigenetic events contribute to prediction error and contextual modulation of drug-seeking behavior.

7. CONSIDERATIONS FOR HISTONE-MEDIATED TREATMENT OF ADDICTION

Although researchers are starting to understand how histone modifications lead to downstream changes in transcription, there are considerable gaps that still need to be understood before complete treatment can be accomplished. The exact mechanism of modification recruitment, coordination, and differences between diseases is largely unknown. These gaps in knowledge limit the potential for therapeutics to target aberrant modifications with spatial and temporal specificity. For example, most HDAC and HDMT inhibitors affect multiple isoforms and other nonhistone proteins with similar activity. This would limit the control that clinicians have over side effects.

At a psychological level, a great deal of attention has been directed toward the idea that HDAC inhibitors may work in a variety of disorders by improving cognitive function. These drugs hold tremendous promise in the treatment of any disorder that involves some type of impaired cognitive function. Research from the rodent laboratory has demonstrated that these drugs are often ineffective on their own—they need to be paired with a behavioral experience, such as extinction. When combined with behavioral interventions, these drugs have great potential to promote treatment outcomes.[137,138] However, this promise comes with potential peril, as a cognitive-enhancing drug could strengthen the impact of an episode of relapse by promoting the transcriptional events associated with the relapse episode. There is therefore a need to administer these drugs under close clinical supervision, during which the clinician has some control over the experiences that the patient has while on the medication.

These basic mechanistic and psychological questions will need to be addressed in future work. What is clear from the literature reviewed in this chapter is that histone modifications play important roles in the development, maintenance, and treatment of addiction. There are many questions that are open and many debates that are unresolved, but there is little doubt that a thorough understanding of the mechanisms of histone modification, transcription, and translation will lead to important scientific discoveries with a high likelihood for clinical translation.

ACKNOWLEDGMENTS

The preparation of this chapter was supported by the National Institutes of Health Grant R01DA025922 (K. M. L.) and T32DA007262 (L. N. H.) and the Department of Defense Grant W81XWH-12-2-0048 (K. M. L.).

REFERENCES

1. Hyman SE, Malenka RC, Nestler EJ. Neural mechanisms of addiction: the role of reward-related learning and memory. *Annu Rev Neurosci*. 2006;29:565–598. http://dx.doi.org/10.1146/annurev.neuro.29.051605.113009.
2. Levenson JM, O'Riordan KJ, Brown KD, Trinh MA, Molfese DL, Sweatt JD. Regulation of histone acetylation during memory formation in the hippocampus. *J Biol Chem*. 2004;279(39):40545–40559. http://dx.doi.org/10.1074/jbc.M402229200.
3. Vecsey CG, Hawk JD, Lattal KM, et al. Histone deacetylase inhibitors enhance memory and synaptic plasticity via CREB:CBP-dependent transcriptional activation. *J Neurosci*. 2007;27:6128–6140. http://dx.doi.org/10.1523/JNEUROSCI.0296-07.2007.
4. Cheung P, Allis CD, Sassone-Corsi P. Signaling to chromatin through histone modifications. *Cell*. 2000;103(2):263–271.
5. Strahl BD, Allis CD. The language of covalent histone modifications. *Nature*. 2000;403:41–45. http://dx.doi.org/10.1038/47412.
6. McGaugh JL. Memory—a century of consolidation. *Science*. 2000;287:248–251. http://dx.doi.org/10.1126/science.287.5451.248.
7. Gould TJ, Leach PT. Cellular, molecular, and genetic substrates underlying the impact of nicotine on learning. *Neurobiol Learn Mem*. 2014;107:108–132. http://dx.doi.org/10.1016/j.nlm.2013.08.004.
8. Wise LE, Varvel SA, Selley DE, et al. delta(9)-Tetrahydrocannabinol-dependent mice undergoing withdrawal display impaired spatial memory. *Psychopharmacology (Berl)*. 2011;217(4):485–494. http://dx.doi.org/10.1007/s00213-011-2305-5.
9. Koob GF, Volkow ND. Neurocircuitry of addiction. *Neuropsychopharmacology*. 2010;35:217–238. http://dx.doi.org/10.1038/npp.2010.4.

10. Marchant NJ, Millan EZ, McNally GP. The hypothalamus and the neurobiology of drug seeking. *Cell Mol Life Sci.* 2012;69(4):581–597. http://dx.doi.org/10.1007/s00018-011-0817-0.
11. Bernardi RE, Ryabinin AE, Berger SP, Lattal KM. Post-retrieval disruption of a cocaine conditioned place preference by systemic and intrabasolateral amygdala beta2- and alpha1-adrenergic antagonists. *Learn Mem.* 2009;16:777–789. http://dx.doi.org/10.1101/lm.1648509.
12. Lalumiere RT, Smith KC, Kalivas PW. Neural circuit competition in cocaine-seeking: roles of the infralimbic cortex and nucleus accumbens shell. *Eur J Neurosci.* 2012;35:614–622. http://dx.doi.org/10.1111/j.1460-9568.2012.07991.x.
13. Robinson TE, Berridge KC. The neural basis of drug craving: an incentive-sensitization theory of addiction. *Brain Res Rev.* 1993;18:247–291. http://dx.doi.org/10.1016/0165-0173(93)90013-P.
14. Siegel S. Classical conditioning, drug tolerance, and drug dependence. In: Smart RG, Glaser FB, Israel Y, Kalant H, Popham RE, eds. *Research Advances in Alcohol and Drug Problems.* 7th ed. Boston, MA: Springer US; 1983:207–246. http://dx.doi.org/10.1007/978-1-4613-3626-6.
15. Tronson NC, Taylor JR. Addiction: a drug-induced disorder of memory reconsolidation. *Curr Opin Neurobiol.* 2013;23(4):573–580. http://dx.doi.org/10.1016/j.conb.2013.01.022.
16. Badiani A, Robinson TE. Drug-induced neurobehavioral plasticity: the role of environmental context. *Behav Pharmacol.* 2004;15:327–339. http://dx.doi.org/10.1097/00008877-200409000-00004.
17. Lattal KM, Radulovic J, Lukowiak K. Extinction: [corrected] does it or doesn't it? The requirement of altered gene activity and new protein synthesis. *Biol Psychiatry.* 2006;60:344–351. http://dx.doi.org/10.1016/j.biopsych.2006.05.038.
18. Peters J, Kalivas PW, Quirk GJ. Extinction circuits for fear and addiction overlap in prefrontal cortex. *Learn Mem.* 2009;16:279–288. http://dx.doi.org/10.1101/lm.1041309.
19. Stefanik MT, Kupchik YM, Brown RM, Kalivas PW. Optogenetic evidence that pallidal projections, not nigral projections, from the nucleus accumbens core are necessary for reinstating cocaine seeking. *J Neurosci.* 2013;33(34):13654–13662. http://dx.doi.org/10.1523/JNEUROSCI.1570-13.2013.
20. Tye KM, Cone JJ, Schairer WW, Janak PH. Amygdala neural encoding of the absence of reward during extinction. *J Neurosci.* 2010;30:116–125. http://dx.doi.org/10.1523/JNEUROSCI.4240-09.2010.
21. Hyman SE. Addiction: a disease of learning and memory. *Am J Psychiatry.* 2005;162:1414–1422. http://dx.doi.org/10.1176/appi.ajp.162.8.1414.
22. Nic Dhonnchadha BÁ, Kantak KM. Cognitive enhancers for facilitating drug cue extinction: insights from animal models. *Pharmacol Biochem Behav.* 2011;99(2):229–244. http://dx.doi.org/10.1016/j.pbb.2011.01.018.
23. Brooks DC. Recent and remote extinction cues reduce spontaneous recovery. *Q J Exp Psychol B.* 2000;53:25–58. http://dx.doi.org/10.1080/027249900392986.
24. Shaham Y, Adamson LK, Grocki S, Corrigall WA. Reinstatement and spontaneous recovery of nicotine seeking in rats. *Psychopharmacology (Berl).* 1997;130:396–403. http://dx.doi.org/10.1007/s002130050256.
25. Crombag HS, Bossert JM, Koya E, Shaham Y. Review. Context-induced relapse to drug seeking: a review. *Philos Trans R Soc Lond B Biol Sci.* 2008;363:3233–3243. http://dx.doi.org/10.1098/rstb.2008.0090.
26. Crombag HS, Shaham Y. Renewal of drug seeking by contextual cues after prolonged extinction in rats. *Behav Neurosci.* 2002;116:169–173. http://dx.doi.org/10.1037/0735-7044.116.1.169.

27. Tipps ME, Raybuck JD, Lattal KM. Substance abuse, memory, and post-traumatic stress disorder. *Neurobiol Learn Mem*. 2014;112C:87–100. http://dx.doi.org/10.1016/j.nlm.2013.12.002.
28. Portela A, Esteller M. Epigenetic modifications and human disease. *Nat Biotechnol*. 2010;28:1057–1068. http://dx.doi.org/10.1038/nbt.1685.
29. Tsankova N, Renthal W, Kumar A, Nestler EJ. Epigenetic regulation in psychiatric disorders. *Neuroscience*. 2007;8:355–367. http://dx.doi.org/10.1038/nrn2132.
30. Sharma S, Kelly TK, Jones PA. Epigenetics in cancer. *Carcinogenesis*. 2009;31(1):27–36. http://dx.doi.org/10.1093/carcin/bgp220.
31. Malvaez M, Barrett RM, Wood MA, Sanchis-Segura C. Epigenetic mechanisms underlying extinction of memory and drug-seeking behavior. *Mamm Genome*. 2009;20:612–623. http://dx.doi.org/10.1007/s00335-009-9224-3.
32. Robison AJ, Nestler EJ. Transcriptional and epigenetic mechanisms of addiction. *Nat Rev Neurosci*. 2011;12(11):623–637. http://dx.doi.org/10.1038/nrn3111.
33. Lv J, Xin Y, Zhou W, Qiu Z. The epigenetic switches for neural development and psychiatric disorders. *J Genet Genomics*. 2013;40(7):339–346. http://dx.doi.org/10.1016/j.jgg.2013.04.007.
34. Darcy MJ, Trouche S, Jin S-X, Feig LA. Ras-GRF2 mediates long-term potentiation, survival, and response to an enriched environment of newborn neurons in the hippocampus. *Hippocampus*. 2014;13:1–13. http://dx.doi.org/10.1002/hipo.22313.
35. Shalin SC, Hernandez CM, Dougherty MK, Morrison DK, Sweatt JD. Kinase suppressor of Ras1 compartmentalizes hippocampal signal transduction and subserves synaptic plasticity and memory formation. *Neuron*. 2006;50(5):765–779. http://dx.doi.org/10.1016/j.neuron.2006.04.029.
36. Cedar H, Bergman Y. Linking DNA methylation and histone modification: patterns and paradigms. *Nat Rev Genet*. 2009;10:295–304. http://dx.doi.org/10.1038/nrg2540.
37. Greer EL, Shi Y. Histone methylation: a dynamic mark in health, disease and inheritance. *Nat Rev Genet*. 2012;13(5):343–357. http://dx.doi.org/10.1038/nrg3173.
38. Smith E, Shilatifard A. The chromatin signaling pathway: diverse mechanisms of recruitment of histone-modifying enzymes and varied biological outcomes. *Mol Cell*. 2010;40(5):689–701. http://dx.doi.org/10.1016/j.molcel.2010.11.031.
39. Braunstein M, Sobel RE, Allis CD, Turner BM, Broach JR. Efficient transcriptional silencing in Saccharomyces cerevisiae requires a heterochromatin histone acetylation pattern. *Mol Cell Biol*. 1996;16(8):4349–4356.
40. Grewal SIS, Jia S. Heterochromatin revisited. *Nat Rev Genet*. 2007;8:35–46. http://dx.doi.org/10.1038/nrg2008.
41. Santos-Rosa H, Schneider R, Bannister AJ, et al. Active genes are tri-methylated at K4 of histone H3. *Nature*. 2002;419(6905):407–411. http://dx.doi.org/10.1038/nature01071.
42. Kouzarides T. Chromatin modifications and their function. *Cell*. 2007;128(4):693–705. http://dx.doi.org/10.1016/j.cell.2007.02.005.
43. Lee JS, Smith E, Shilatifard A. The language of histone crosstalk. *Cell*. 2010;142(5):682–685. http://dx.doi.org/10.1016/j.cell.2010.08.011.
44. Muratani M, Tansey WP. How the ubiquitin–proteasome system controls transcription. *Nat Rev Mol Cell Biol*. 2003;4(3):192–201. http://dx.doi.org/10.1038/nrm1049.
45. Zhang H, Gao L, Anandhakumar J, Gross DS. Uncoupling transcription from covalent histone modification. *PLoS Genet*. 2014;10:e1004202. http://dx.doi.org/10.1371/journal.pgen.1004202.
46. Wang Y, Jia S. Degrees make all the difference: the multifunctionality of histone H4 lysine 20 methylation. *Epigenetics*. 2009;4(5):273–276. http://dx.doi.org/10.4161/epi.4.5.9212.

47. Maze I, Feng J, Wilkinson MB, Sun H, Shen L, Nestler EJ. Cocaine dynamically regulates heterochromatin and repetitive element unsilencing in nucleus accumbens. *Proc Natl Acad Sci USA*. 2011;108(7):3035–3040. http://dx.doi.org/10.1073/pnas.1015483108.
48. Mozzetta C, Pontis J, Fritsch L, et al. The histone H3 lysine 9 methyltransferases G9a and GLP regulate polycomb repressive complex 2-mediated gene silencing. *Mol Cell*. 2014;53(2):277–289. http://dx.doi.org/10.1016/j.molcel.2013.12.005.
49. Alvaro-Bartolome M, Garcia-Sevilla JA. Dysregulation of cannabinoid CB1 receptor and associated signaling networks in brains of cocaine addicts and cocaine-treated rodents. *Neuroscience*. 2013;247:294–308. http://dx.doi.org/10.1016/j.neuroscience.2013.05.035.
50. Castellano-Pozo M, Santos-Pereira JM, Rondón AG, et al. R loops are linked to histone H3 S10 phosphorylation and chromatin condensation. *Mol Cell*. 2013;52(4): 583–590. http://dx.doi.org/10.1016/j.molcel.2013.10.006.
51. Johansen KM, Johansen J. Regulation of chromatin structure by histone H3S10 phosphorylation. *Chromosome Res*. 2006;14(4):393–404. http://dx.doi.org/10.1007/s10577-006-1063-4.
52. Wilkins BJ, Rall NA, Ostwal Y, et al. A cascade of histone modifications induces chromatin condensation in mitosis. *Science*. 2014;343(6166):77–80. http://dx.doi.org/10.1126/science.1244508.
53. Zhang Y, Griffin K, Mondal N, Parvin JD. Phosphorylation of histone H2A inhibits transcription on chromatin templates. *J Biol Chem*. 2004;279(21):21866–21872. http://dx.doi.org/10.1074/jbc.M400099200.
54. Chen F, Kan H, Castranova V. Methylation of lysine 9 of histone H3: role of heterochromatin modulation and tumorigenesis. In: *Handbook of Epigenetics: The Molecular and Medical Genetics*. 1st ed. Amsterdam: Elsevier; 2011:149–157. http://dx.doi.org/10.1016/B978-0-12-375709-8.00010-1.
55. Sims JK, Magazinnik-Spektor T, Houston SI, Wu S, Rice JC. Histone modifications and epigenetics. In: Tost J, ed. *Epigenetics*. Norfolk: Horizon Scientific Press; 2008:407.
56. Aguilar-Valles A, Vaissière T, Griggs EM, et al. Methamphetamine-associated memory is regulated by a writer and an eraser of permissive histone methylation. *Biol Psychiatry*. 2014;76(1):57–65. http://dx.doi.org/10.1016/j.biopsych.2013.09.014.
57. Jenuwein T, Allis CD. Translating the histone code. *Science*. 2001;293:1074–1080. http://dx.doi.org/10.1126/science.1063127.
58. Loidl P. Histone acetylation: facts and questions. *Chromosoma*. 1994;103:441–449. http://dx.doi.org/10.1007/s004120050053.
59. Dokmanovic M, Clarke C, Marks PA. Histone deacetylase inhibitors: overview and perspectives. *Mol Cancer Res*. 2007;5:981–989. http://dx.doi.org/10.1158/1541-7786.MCR-07-0324.
60. Dietrich J-B, Takemori H, Grosch-Dirrig S, Bertorello A, Zwiller J. Cocaine induces the expression of MEF2C transcription factor in rat striatum through activation of SIK1 and phosphorylation of the histone deacetylase HDAC5. *Synapse*. 2012;66(1):61–70. http://dx.doi.org/10.1002/syn.20988.
61. Brami-Cherrier K, Roze E, Girault JA, Betuing S, Caboche J. Role of the ERK/MSK1 signaling pathway in chromatin remodelling and brain responses to drugs of abuse. *J Neurochem*. 2009;108(6):1323–1335. http://dx.doi.org/10.1111/j.1471-4159.2009.05879.x.
62. Clayton AL, Rose S, Barratt MJ, Mahadevan LC. Phosphoacetylation of histone H3 on c-fos- and c-jun-associated nucleosomes upon gene activation. *EMBO J*. 2000;19(14): 3714–3726.
63. Kim J-Y, Banerjee T, Vinckevicius A, et al. A role for WDR5 in integrating threonine 11 phosphorylation to lysine 4 methylation on histone H3 during androgen signaling and in prostate cancer. *Mol Cell*. 2014;54:1–13. http://dx.doi.org/10.1016/j.molcel.2014.03.043.

64. Weake VM, Workman JL. Histone ubiquitination: triggering gene activity. *Mol Cell.* 2008;29:653–663. http://dx.doi.org/10.1016/j.molcel.2008.02.014.
65. Welchman RL, Gordon C, Mayer RJ. Ubiquitin and ubiquitin-like proteins as multifunctional signals. *Nat Rev Mol Cell Biol.* 2005;6:599–609. http://dx.doi.org/10.1038/nrm1700.
66. Wilkinson KA, Henley JM. Mechanisms, regulation and consequences of protein SUMOylation. *Biochem J.* 2012;428(2):133–145. http://dx.doi.org/10.1042/BJ20100158.
67. Tulin A, Chinenov Y, Spradling A. Regulation of chromatin structure and gene activity by poly(ADP-ribose) polymerases. *Curr Top Dev Biol.* 2003;56:55–83.
68. Liu R, Lei JX, Luo C, et al. Increased EID1 nuclear translocation impairs synaptic plasticity and memory function associated with pathogenesis of Alzheimer's disease. *Neurobiol Dis.* 2012;45(3):902–912. http://dx.doi.org/10.1016/j.nbd.2011.12.007.
69. Salles A, Romano A, Freudenthal R. Synaptic NF-kappa B pathway in neuronal plasticity and memory. *J Physiol Paris.* 2014;50, in press. http://dx.doi.org/10.1016/j.jphysparis.2014.05.002.
70. De Vos M, El Ramy R, Quénet D, et al. Poly(ADP-ribose) polymerase 1 (PARP1) associates with E3 ubiquitine-protein ligase UHRF1 and modulates UHRF1 biological functions. *J Biol Chem.* 2014;1(23):16223–16238. http://dx.doi.org/10.1074/jbc.M113.527424.
71. Chen WG, Chang Q, Lin Y, et al. Derepression of BDNF transcription involves calcium-dependent phosphorylation of MeCP2. *Science.* 2003;302(5646):885–889. http://dx.doi.org/10.1126/science.1086446.
72. Stephens DN, Crombag HS, Duka T. The challenge of studying parallel behaviors in humans and animal models. *Curr Top Behav Neurosci.* 2013;13:611–645. http://dx.doi.org/10.1007/7854_2011_133.
73. Bohacek J, Mansuy IM. Epigenetic inheritance of disease and disease risk. *Neuropsychopharmacology.* 2013;38(1):220–236. http://dx.doi.org/10.1038/npp.2012.110.
74. Petronis A. Epigenetics as a unifying principle in the aetiology of complex traits and diseases. *Nature.* 2010;465:721–727. http://dx.doi.org/10.1038/nature09230.
75. Renthal W, Nestler EJ. Histone acetylation in drug addiction. *Semin Cell Dev Biol.* 2009;20(4):387–394. http://dx.doi.org/10.1016/j.semcdb.2009.01.005.
76. Maze I, Covington HE, Dietz DM, et al. Essential role of the histone methyltransferase G9a in cocaine-induced plasticity. *Science.* 2010;327(5962):213–216. http://dx.doi.org/10.1126/science.1179438.
77. Renthal W, Carle TL, Maze I, et al. Delta FosB mediates epigenetic desensitization of the c-fos gene after chronic amphetamine exposure. *J Neurosci.* 2008;28(29):7344–7349. http://dx.doi.org/10.1523/JNEUROSCI.1043-08.2008.
78. Sun H, Maze I, Dietz DM, et al. Morphine epigenomically regulates behavior through alterations in histone H3 lysine 9 dimethylation in the nucleus accumbens. *J Neurosci.* 2012;32(48):17454–17464. http://dx.doi.org/10.1523/JNEUROSCI.1357-12.2012.
79. Chase KA, Sharma RP. Nicotine induces chromatin remodelling through decreases in the methyltransferases GLP, G9a, Setdb1 and levels of H3K9me2. *Int J Neuropsychopharmacol.* 2012;2:1–10. http://dx.doi.org/10.1017/S1461145712001101.
80. McQuown SC, Wood MA. HDAC3 and the molecular brake pad hypothesis. *Neurobiol Learn Mem.* 2011;96(1):27–34. http://dx.doi.org/10.1016/j.nlm.2011.04.005.
81. Covington HE, Maze I, Sun H, et al. A role for repressive histone methylation in cocaine-induced vulnerability to stress. *Neuron.* 2011;71(4):656–670. http://dx.doi.org/10.1016/j.neuron.2011.06.007.
82. Toth I, Neumann ID. Animal models of social avoidance and social fear. *Cell Tissue Res.* 2013;354(1):107–118. http://dx.doi.org/10.1007/s00441-013-1636-4.

83. Bahi A, Boyer F, Dreyer JL. Role of accumbens BDNF and TrkB in cocaine-induced psychomotor sensitization, conditioned-place preference, and reinstatement in rats. *Psychopharmacology (Berl)*. 2008;199(2):169–182. http://dx.doi.org/10.1007/s00213-008-1164-1.
84. Corominas-Roso M, Roncero C, Eiroa-Orosa FJ, et al. Brain-derived neurotrophic factor serum levels in cocaine-dependent patients during early abstinence. *Eur Neuropsychopharmacol*. 2013;23(9):1078–1084. http://dx.doi.org/10.1016/j.euroneuro.2012.08.016.
85. Fuchikami M, Yamamoto S, Morinobu S, Takei S, Yamawaki S. Epigenetic regulation of BDNF gene in response to stress. *Psychiatry Investig*. 2010;7(4):251–256. http://dx.doi.org/10.4306/pi.2010.7.4.251.
86. Kennedy PJ, Feng J, Robison AJ, et al. Class I HDAC inhibition blocks cocaine-induced plasticity by targeted changes in histone methylation. *Nat Neurosci*. 2013;16(4):434–440. http://dx.doi.org/10.1038/nn.3354.
87. Sen N. Epigenetic regulation of memory by acetylation and methylation of chromatin: implications in neurological disorders, aging, and addiction. *Neuromol Med*. 2014, in press. http://dx.doi.org/10.1007/s12017-014-8306-x.
88. Chavda B, Arnott JA, Planey SL. Targeting protein palmitoylation: selective inhibitors and implications in disease. *Exp Opin Drug Discov*. 2014;9:1005–1019. http://dx.doi.org/10.1517/17460441.2014.933802.
89. Dieker J, Muller S. Epigenetic histone code and autoimmunity. *Clin Rev Allergy Immunol*. 2010;39(1):78–84. http://dx.doi.org/10.1007/s12016-009-8173-7.
90. Khanal P, Kim G, Lim S-C, et al. Prolyl isomerase Pin1 negatively regulates the stability of SUV39H1 to promote tumorigenesis in breast cancer. *FASEB J*. 2013;27(11):4606–4618. http://dx.doi.org/10.1096/fj.13-236851.
91. Wagner JM, Hackanson B, Lübbert M, Jung M. Histone deacetylase (HDAC) inhibitors in recent clinical trials for cancer therapy. *Clin Epigenet*. 2010;1:117–136. http://dx.doi.org/10.1007/s13148-010-0012-4.
92. Yoshida M, Hoshikawa Y, Koseki K, Mori K, Beppu T. Structural specificity for biological activity of trichostatin A, a specific inhibitor of mammalian cell cycle with potent differentiation-inducing activity in Friend leukemia cells. *J Antibiot*. 1990;43(9):1101–1106.
93. Zhang J, Zhong Q. Histone deacetylase inhibitors and cell death. *Cell Mol Life Sci*. 2014;70(20):3885–3901. http://dx.doi.org/10.1007/s00018-014-1656-6.
94. Drummond DC, Noble CO, Kirpotin DB, Guo Z, Scott GK, Benz CC. Clinical development of histone deacetylase inhibitors as anticancer agents. *Annu Rev Pharmacol Toxicol*. 2005;45:495–528. http://dx.doi.org/10.1146/annurev.pharmtox.45.120403.095825.
95. Romieu P, Host L, Gobaille S, Sandner G, Aunis D, Zwiller J. Histone deacetylase inhibitors decrease cocaine but not sucrose self-administration in rats. *J Neurosci*. 2008;28(38):9342–9348. http://dx.doi.org/10.1523/JNEUROSCI.0379-08.2008.
96. Arora DS, Nimitvilai S, Teppen TL, et al. Hyposensitivity to gamma-aminobutyric acid in the ventral tegmental area during alcohol withdrawal: reversal by histone deacetylase inhibitors. *Neuropsychopharmacology*. 2013;38(9):1674–1684. http://dx.doi.org/10.1038/npp.2013.65.
97. Raybuck JD, McCleery EJ, Cunningham CL, Wood MA, Lattal KM. The histone deacetylase inhibitor sodium butyrate modulates acquisition and extinction of cocaine-induced conditioned place preference. *Pharmacol Biochem Behav*. 2013;106:109–116. http://dx.doi.org/10.1016/j.pbb.2013.02.009.
98. Malvaez M, McQuown SC, Rogge GA, et al. HDAC3-selective inhibitor enhances extinction of cocaine-seeking behavior in a persistent manner. *Proc Natl Acad Sci USA*. 2013;110:2647–2652. http://dx.doi.org/10.1073/pnas.1213364110.

99. Malvaez M, Sanchis-Segura C, Vo D, Lattal KM, Wood MA. Modulation of chromatin modification facilitates extinction of cocaine-induced conditioned place preference. *Biol Psychiatry*. 2010;67:36–43. http://dx.doi.org/10.1016/j.biopsych.2009.07.032.
100. Kumar A, Choi KH, Renthal W, et al. Chromatin remodeling is a key mechanism underlying cocaine-induced plasticity in striatum. *Neuron*. 2005;48(2):303–314. http://dx.doi.org/10.1016/j.neuron.2005.09.023.
101. Renthal W, Kumar A, Xiao G, et al. Genome-wide analysis of chromatin regulation by cocaine reveals a role for sirtuins. *Neuron*. 2009;62(3):335–348. http://dx.doi.org/10.1016/j.neuron.2009.03.026.
102. Schroeder FA, Penta KL, Matevossian A, et al. Drug-induced activation of dopamine D(1) receptor signaling and inhibition of class I/II histone deacetylase induce chromatin remodeling in reward circuitry and modulate cocaine-related behaviors. *Neuropsychopharmacology*. 2008;33:2981–2992. http://dx.doi.org/10.1038/npp.2008.15.
103. Sun J, Wang L, Jiang B, Hui B, Lv Z, Ma L. The effects of sodium butyrate, an inhibitor of histone deacetylase, on the cocaine- and sucrose-maintained self-administration in rats. *Neurosci Lett*. 2008;441(1):72–76. http://dx.doi.org/10.1016/j.neulet.2008.05.010.
104. Wang Y, Krishnan HR, Ghezzi A, Yin JCP, Atkinson NS. Drug-induced epigenetic changes produce drug tolerance. *PLoS Biol*. 2007;5:2342–2353. http://dx.doi.org/10.1371/journal.pbio.0050265.
105. Zou JY, Crews FT. Release of neuronal HMGB1 by ethanol through decreased HDAC activity activates brain neuroimmune signaling. *PLoS One*. 2014;9(2):e87915. http://dx.doi.org/10.1371/journal.pone.0087915.
106. Renthal W, Maze I, Krishnan V, et al. Histone deacetylase 5 epigenetically controls behavioral adaptations to chronic emotional stimuli. *Neuron*. 2007;56(3):517–529. http://dx.doi.org/10.1016/j.neuron.2007.09.032.
107. Maze I, Nestler EJ. The epigenetic landscape of addiction. *Ann N Y Acad Sci*. 2011;1216:99–113. http://dx.doi.org/10.1111/j.1749-6632.2010.05893.x.
108. Tsankova NM, Kumar A, Nestler EJ. Histone modifications at gene promoter regions in rat hippocampus after acute and chronic electroconvulsive seizures. *J Neurosci*. 2004;24(24):5603–5610. http://dx.doi.org/10.1523/JNEUROSCI.0589-04.2004.
109. Sakharkar AJ, Tang L, Zhang H, Chen Y, Grayson DR, Pandey SC. Effects of acute ethanol exposure on anxiety measures and epigenetic modifiers in the extended amygdala of adolescent rats. *Int J Neuropsychopharmacol*. 2014, in press. http://dx.doi.org/10.1017/S1461145714001047.
110. Shepard BD, Joseph RA, Kannarkat GT, Rutledge TM, Tuma DJ, Tuma PL. Alcohol-induced alterations in hepatic microtubule dynamics can be explained by impaired histone deacetylase 6 function. *Hepatology*. 2008;48(5):1671–1679. http://dx.doi.org/10.1002/hep.22481.
111. Starkman B, Sakharkar AJ, Pandey S. Epigenetics—beyond the genome in alcoholism. *Alcohol Res*. 2012;34(3):293–305.
112. Gruber AJ, McDonald RJ. Context, emotion, and the strategic pursuit of goals: interactions among multiple brain systems controlling motivated behavior. *Front Behav Neurosci*. 2012;6:50. http://dx.doi.org/10.3389/fnbeh.2012.00050.
113. Pandey SC, Ugale R, Zhang H, Tang L, Prakash A. Brain chromatin remodeling: a novel mechanism of alcoholism. *J Neurosci*. 2008;28(14):3729–3737. http://dx.doi.org/10.1523/JNEUROSCI.5731-07.2008.
114. Sakharkar AJ, Zhang H, Tang L, Shi G, Pandey SC. Histone deacetylases (HDAC)-induced histone modifications in the amygdala: a role in rapid tolerance to the anxiolytic effects of ethanol. *Alcohol Clin Exp Res*. 2012;36(1):61–71. http://dx.doi.org/10.1111/j.1530-0277.2011.01581.x.
115. Li J, Guo Y, Schroeder FA, et al. Dopamine D2-like antagonists induce chromatin remodeling in striatal neurons through cyclic AMP-protein kinase A and NMDA

receptor signaling. *J Neurochem*. 2004;90(5):1117–1131. http://dx.doi.org/10.1111/j.1471-4159.2004.02569.x.
116. Qiang M, Denny A, Lieu M, Carreon S, Li J. Histone H3K9 modifications are a local chromatin event involved in ethanol-induced neuroadaptation of the NR2B gene. *Epigenetics*. 2011;6(9):1095–1104. http://dx.doi.org/10.4161/epi.6.9.16924.
117. Sheng J, Lv ZG, Wang L, Zhou Y, Hui B. Histone H3 phosphoacetylation is critical for heroin-induced place preference. *Neuroreport*. 2011;22(12):575–580. http://dx.doi.org/10.1097/WNR.0b013e328348e6aa.
118. Adenuga D, Yao H, March TH, Seagrave J, Rahman I. Histone deacetylase 2 is phosphorylated, ubiquitinated, and degraded by cigarette smoke. *Am J Respir Cell Mol Biol*. 2009;40:464–473. http://dx.doi.org/10.1165/rcmb.2008-0255OC.
119. Schmitt KC, Reith MEA. Regulation of the dopamine transporter: aspects relevant to psychostimulant drugs of abuse. *Ann N Y Acad Sci*. 2010;1187:316–340. http://dx.doi.org/10.1111/j.1749-6632.2009.05148.x.
120. Kannan P, Yu Y, Wankhade S, Tainsky MA. PolyADP-ribose polymerase is a coactivator for AP-2-mediated transcriptional activation. *Nucleic Acids Res*. 1999;27(3):866–874.
121. Biliński P, Wojtyła A, Kapka-Skrzypczak L, Chwedorowicz R, Cyranka M, Studziński T. Epigenetic regulation in drug addiction. *Ann Agric Environ Med*. 2012;19(3):491–496.
122. Henikoff S, Shilatifard A. Histone modification: cause or cog? *Trends Genet*. 2011;27(10):389–396. http://dx.doi.org/10.1016/j.tig.2011.06.006.
123. Nakamura Y, Umehara T, Nakano K, et al. Crystal structure of the human BRD2 bromodomain: insights into dimerization and recognition of acetylated histone H4. *J Biol Chem*. 2007;282:4193–4201. http://dx.doi.org/10.1074/jbc.M605971200.
124. Brumfiel G. Disputed definitions. *Nature*. 2008;455:1023–1028.
125. Hammoud SS, Nix DA, Zhang H, Purwar J, Carrell DT, Cairns BR. Distinctive chromatin in human sperm packages genes for embryo development. *Nature*. 2009;460:473–478. http://dx.doi.org/10.1038/nature08162.
126. Heard E, Martienssen RA. Transgenerational epigenetic inheritance: myths and mechanisms. *Cell*. 2014;157(1):95–109. http://dx.doi.org/10.1016/j.cell.2014.02.045.
127. Vassoler FM, Johnson-collins NL, Carini LM, Byrnes EM. Next generation effects of female adolescent morphine exposure: sex-specific alterations in response to acute morphine emerge before puberty. *Behav Pharmacol*. 2011;25(2):173–181. http://dx.doi.org/10.1097/FBP.0000000000000032.
128. Vassoler FM, White SL, Schmidt HD, Sadri-Vakili G, Pierce RC. Epigenetic inheritance of a cocaine-resistance phenotype. *Nat Neurosci*. 2013;16:42–47. http://dx.doi.org/10.1038/nn.3280.
129. Norrholm SD, Jovanovic T, Smith AK, et al. Differential genetic and epigenetic regulation of catechol-O-methyltransferase is associated with impaired fear inhibition in posttraumatic stress disorder. *Front Behav Neurosci*. 2013;7:30. http://dx.doi.org/10.3389/fnbeh.2013.00030.
130. Bilder RM, Volavka J, Lachman HM, Grace AA. The catechol-O-methyltransferase polymorphism: relations to the tonic-phasic dopamine hypothesis and neuropsychiatric phenotypes. *Neuropsychopharmacology*. 2004;29(11):1943–1961. http://dx.doi.org/10.1038/sj.npp.1300542.
131. Matsumoto M, Weickert CS, Akil M, et al. Catechol O-methyltransferase mRNA expression in human and rat brain: evidence for a role in cortical neuronal function. *Neuroscience*. 2003;116(1):127–137. http://dx.doi.org/10.1016/S0306-4522(02)00556-0.
132. Dias BG, Ressler KJ. Parental olfactory experience influences behavior and neural structure in subsequent generations. *Nat Neurosci*. 2014;17(1):89–96.

133. Lattal KM, Wood MA. Epigenetics and persistent memory: implications for reconsolidation and silent extinction beyond the zero. *Nat Neurosci.* 2013; 16:124–129. http://dx.doi.org/10.1038/nn.3302.
134. Stafford JM, Lattal KM. Is an epigenetic switch the key to persistent extinction? *Neurobiol Learn Mem.* 2011;96:35–40. http://dx.doi.org/10.1016/j.nlm.2011.04.012.
135. Stafford JM, Raybuck JD, Ryabinin AE, Lattal KM. Increasing histone acetylation in the hippocampus-infralimbic network enhances fear extinction. *Biol Psychiatry.* 2012;72:25–33. http://dx.doi.org/10.1016/j.biopsych.2011.12.012.
136. Schultz W. Behavioral dopamine signals. *Trends Neurosci.* 2007;30:203–210. http://dx.doi.org/10.1016/j.tins.2007.03.007.
137. Davis M, Myers KM, Chhatwal J, Ressler KJ. Pharmacological treatments that facilitate extinction of fear: relevance to psychotherapy. *NeuroRx.* 2006;3(1):82–96. http://dx.doi.org/10.1016/j.nurx.2005.12.008.
138. Kiefer F, Dinter C. New approaches to addiction treatment based on learning and memory. *Curr Top Behav Neurosci.* 2013;13:671–684. http://dx.doi.org/10.1007/7854_2011_147.
139. Jones PL, Veenstra GJ, Wade PA, et al. Methylated DNA and MeCP2 recruit histone deacetylase to repress transcription. *Nat Neurosci.* 1998;19(2):187–191. http://dx.doi.org/10.1038/561.
140. Barski A, Cuddapah S, Cui K, et al. High-resolution profiling of histone methylations in the human genome. *Cell.* 2007;129(4):823–837. http://dx.doi.org/10.1016/j.cell.2007.05.009.
141. Vassoler FM, Sadri-Vakili G. Mechanisms of transgenerational inheritance of addictive-like behaviors. *Neuroscience.* 2014;264:198–206. http://dx.doi.org/10.1016/j.neuroscience.2013.07.064.

CHAPTER FOUR

Modeling the Molecular Epigenetic Profile of Psychosis in Prenatally Stressed Mice

Alessandro Guidotti, Erbo Dong, Patricia Tueting, Dennis R. Grayson
The Psychiatric Institute, Department of Psychiatry, College of Medicine, University of Illinois at Chicago, Chicago, Illinois, USA

Contents

Abstract

Based on postmortem brain studies, our overarching epigenetic hypothesis is that chronic schizophrenia (SZ) is a psychopathological condition involving dysregulation of the dynamic equilibrium among DNA methylation/demethylation network components and the expression of SZ target genes, including GABAergic and glutamatergic genes.

SZ has a natural course, starting with a prodromal phase, a first episode that occurs in adolescents or in young adults, and later deterioration over the adult years. Hence, the epigenetic status at each neurodevelopmental stage of the disease cannot be studied just in postmortem brain of chronic SZ patients, but requires the use of neurodevelopmental animal models. We have directed the focus of our research toward studying the epigenetic signature of the SZ brain in the offspring of dams stressed during pregnancy (PRS mice). Adult PRS mice have behavioral deficits reminiscent of behaviors observed in psychotic patients. The adult PRS brain, like that of postmortem chronic SZ patients, is characterized by a significant increase in DNA methyltransferase 1, Tet methylcytosine dioxygenase 1 (TET1), 5-methylcytosine, and 5-hydroxymethylcytosine at SZ candidate gene promoters and a reduction in the expression of glutamatergic and

ISSN 1877-1173
http://dx.doi.org/10.1016/B978-0-12-800977-2.00004-8

GABAergic genes. In PRS mice, measurements of epigenetic biomarkers for SZ can be assessed at different stages of development with the goal of further elucidating the pathophysiology of this disease and predicting treatment responses at specific stages of the illness, with particular attention to early detection and possibly early intervention.

1. THE EPIGENETIC HYPOTHESIS OF PSYCHOSIS

There are several epidemiological, clinical, and molecular peculiarities associated with major psychoses [schizophrenia (SZ) and bipolar (BP) disorder] that are difficult to reconcile with a Mendelian genetic disorder and in contrast correspond to features of an altered epigenetic homeostasis. Such features include (1) incomplete phenotypic concordance between monozygotic twins (only about 50% concordance), (2) fluctuating disease course with periods of remission and relapse, (3) peaks of susceptibility to disease coinciding with major hormonal changes, and (4) parent-of-origin effects.[1,2]

1.1. DNA methyltransferase

In support of a role for aberrant epigenetic mechanisms in the pathogenesis of SZ and BP disorders, it recently has been reported that downregulation of glutamic acid decarboxylase 67 (gene symbol = *GAD1*) and reelin (*RELN*) in GABAergic neurons[3–8] and of brain-derived nerve growth factor (*BDNF*)[9–11] and vesicular glutamate transporter (VGLUT1) in glutamatergic neurons[12] is associated with an overexpression of DNA methyltransferase (DNMT) 1 and 3a in the cortical BA9, 10, and 17 and in the striatum of the SZ and BP postmortem brains.[13–17] DNMT belongs to a family of enzymes that catalyze the transfer of a methyl group from *S*-adenosylmethionine (SAM) to the 5 carbon of cytosine. DNMTs are highly expressed in telencephalic GABAergic interneurons in both humans and rodents.[13,18] In addition to the increased expression of DNMTs, the hypothesis that an epigenetic DNA methylation pathology operates in the transcriptional downregulation of several target genes in SZ and BP disorder patients is supported by the following evidence: (1) increased SAM levels[19]; (2) enrichment of 5-methylcytosine (5MC) and 5-hydroxymethylcytosine (5HMC) at *RELN*,[20–22] *BDNF*,[11,23] and *GAD1* promoters[24–28]; (3) increased histone methylation at GABAergic gene promoters[29]; (4) an inverse correlation between DNA methylation of the *BDNF*, *RELN*, *and GAD1* genes and the level of their expression in the PFC[23,28]; and

(5) the evidence of epigenetic dysregulation of several other GABAergic and glutamatergic genes.[30]

Support for the hypothesis that a chromatin methylation pathology is a major contributor to the downregulation of GABAergic and glutamatergic genes in psychotic patients is sustained by clinical studies conducted in the early 1970s.[31] In these studies, methionine, the precursor of SAM, when administered to SZ patients in large doses (10/20 g/day for 3–4 weeks), was reported to exacerbate psychotic symptoms. In both mouse frontal cortex (FC) and neuronal cultures, the administration of large doses of methionine induces an increase in SAM and hypermethylation of selective CpG-rich promoters, including *GAD1* and *RELN*, and facilitates downregulation of their expression.[24–27,32,33] Importantly, brain levels of GAD65 (GAD2) and the housekeeping genes are not affected.

These data are consistent with the epigenetic theory of major psychosis[34] and suggest that DNA methylation and DNA demethylation associated with GABAergic and glutamatergic gene regulatory domains are important casual events in the pathogenesis of SZ and BP disorders.

1.2. DNA demethylase

DNA methylation, potentially the longest-lasting epigenetic mark, is uniquely able to account for the chronicity and often intractable nature of SZ and BP disorders. Recent evidence suggests that for inducible genes, steady-state levels of DNA methylation are the result of a dynamic equilibrium between the counterbalancing actions of DNMTs, referred to as "DNA writers" because they modify DNA by adding methyl groups to cytosines, and an active DNA demethylation pathway (cytosine deaminase and base excision repair [BER] pathway), referred to as "DNA erasers" because they remove methyl groups from cytosines.[35]

BER–DNA demethylation is thought to occur as a result of the coordinated actions of ten-eleven translocation (TET) proteins, which hydroxylate 5MC to form 5HMC. Then, through a growth arrest and DNA damage (GADD45) protein coordinated process, APOBEC cytidine deaminases convert 5HMC to 5-hydroxymethyuracil (5HMU), which can be excised by thymine glycosylases leading to the restoration of the nonmethylated state (Fig. 4.1).

It is generally believed that the role of TET is to facilitate the removal of 5MC via formation of the intermediate 5HMC.[35] Recent studies of the postmortem brain by Dong *et al.*[28] indicate that there is an almost twofold

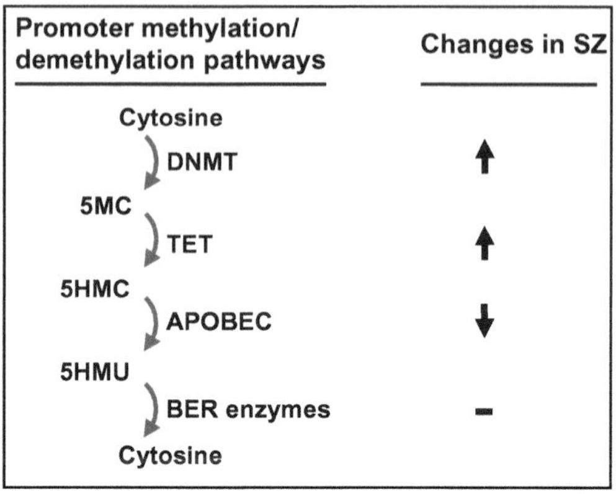

Figure 4.1 Promoter methylation/demethylation pathways; Changes in SZ. C, cytosine; DNMT, DNA methyltransferase; 5MC, 5-methylcytosine; TET, ten-eleven translocation protein; 5HMC, 5-hydroxymethylcytosine; APOBEC, deaminase apolipoprotein B RNA editing; 5HMU, 5-hydroxymethyluracil; BER, base excision repair.

increase in TET1 mRNA and protein in the inferior parietal lobule of psychotic patients. Consistent with this increase in TET1, levels of 5HMC in total DNA are elevated. Moreover, higher 5HMC levels are detected at *GAD1* and *BDNF-IX* promoters only in the psychotic group. The increase in TET1 in psychotic patients is inversely correlated with a decrease in GAD1 and BDNF-IX mRNA expression.[23,28] In a recent study, early-life maternal deprivation was found to be associated with changes in DNA hydroxymethylation levels in the promoters of genes related to neurological or psychiatric disorders.[36] Although TET-dependent active DNA demethylation and increased gene expression may be operative in the brain under normal physiological conditions,[35] in the brain of psychotic patients, the increase in TET1 positively correlates with an increase in 5HMC at genomic DNA and at promoters including *GAD1* and *BDNF*, and TET upregulation has been associated with a downregulation in the expression of these target genes.[23,28] A possible explanation for the different actions of TET in normal physiology and in psychopathology is that the increase in TET in the psychotic brain is associated with a downregulation of the main APOBEC-deaminating enzymes. Dong *et al.*[28] showed that the most abundant APOBEC isoforms (3A and 3C) are reduced significantly in psychosis (Table 4.1). This reduction in APOBEC enzymes prevents active demethylation of critical promoters resulting in an enrichment (accumulation) of 5HMC, which by recruiting MeCP2 or MBD3–NURD complex (collectively referred to as "DNA readers") acts to suppress transcription.[2]

Table 4.1 Comparison of molecular and behavioral abnormalities in SZ and BP disorder patients and PRS mice

	SZ+BP disorder patients	PRS mice
Molecular changes		
GABAergic gene expression[3–8]	↓	↓
DNMT1, 3A and TET1 expression[13–17,28]	↑	↑
5MC and 5HMC levels at *GAD1*, *RELN*, and *BDNF* promoters[11,23,28]	↑	↑
APOBEC 3A/3C expression[28]	↓	ND
Behavioral changes[37–39]		
Positive symptoms (stereotype behavior, NMDA receptor antagonist increased sensitivity)	↑	↑
Negative symptoms (social interactions)	↓	↓
Cognitive, information processing deficit (PPI, fear conditioning)	↑	↑

Molecular changes refer to PFC and hippocampus.

Patients with SZ or BP disorders often receive antipsychotic medications. Since control subjects rarely receive these medications, the question is whether the altered expression of GABAergic and glutamatergic gene expression in the brain of these patients is the consequence of protracted antipsychotic treatment rather than an etiopathogenetic signature of SZ or BP disorder. Although several postmortem studies have reported no correlation between the levels of GAD1, RELN, or BDNF mRNA and proteins and lifetime dosage of antipsychotic medications,[4,40–42] the low statistical power of these postmortem studies may not be sufficient to definitively rule out medication as a confounding variable.

Therefore, we and others have turned to animal studies to address possible medication issues. Protracted haloperidol treatment of rats, at least in one study,[4] fails to change RELN mRNA content in the cortex and cerebellum. In another study, it was shown that protracted haloperidol treatment fails to change the expression of GAD1 mRNA in the PFC of nonhuman primates.[40] Additionally, it has been reported that chronic (27 days) haloperidol or clozapine treatment increases, rather than decreases, the expression of GAD1 in corticolimbic structures.[41] Fatemi *et al.*[42] also reported that chronic olanzapine administration facilitates the differential expression of genes involved in signal transduction, cell communication, metabolism, and immune responses and leads to an upregulation of RELN expression in the FC of rats. Finally, Costa and his group[43] had previously reported that the turnover rate of GABA fails to change with haloperidol and increases with clozapine treatment in rats. Collectively, these data suggest that the downregulation of RELN, BDNF, and GAD1 in the brains of SZ and BP disorder patients is not the consequence of antipsychotic treatment. Nevertheless, more extensive studies involving additional typical and atypical antipsychotic administration are needed.

2. PRENATAL OR EARLY-LIFE STRESS AND IMPAIRED EPIGENETIC PROFILE

Clinical studies have shown that exposure of pregnant women to psychological stress, malnutrition, or viral infection exerts profound effects on neurodevelopment and the behavior of their children. Moreover, prenatal stress is associated with an increased incidence of SZ in these children later in life.[37,44–48] It has been suggested that prenatal or early-life stress, through altered epigenetic mechanisms, is a predisposing factor for SZ and BP disorders by disrupting time- and spatial-dependent neurodevelopmental cues associated with neuronal differentiation and synaptic pruning.[8,49–52] Thus, prenatal stress is considered to be a contributing factor for several neurodevelopmental disorders, including psychotic disorders.[52]

Stevens and colleagues[53,54] demonstrated in mice born to stressed pregnant mothers (here defined as PRS mice) that the migration of inhibitory neuronal progenitors from the medial ganglionic eminence to the cortex is delayed and that this delay persists over time resulting in a reduction of GAD1-positive neurons in the neonatal medial FC of PRS mice. We have also found that the levels of GAD1 in GABAergic neurons in the cortex and hippocampus are altered by prenatal stress (Erbo Dong, unpublished data).[38,39]

Uchida *et al.*,[55] by administering 5-bromo-2′-deoxyuridine to label progenitor neurons, found that the number of GABAergic neurons, but not cortical plate cells, was significantly decreased in the mouse fetal brain during maternal stress. Postnatally, the density of parvalbumin-positive GABAergic neurons was significantly decreased in the PFC, hippocampus, and sensorimotor cortex, suggesting that prenatal stress, in addition to downregulating the expression of GAD1, results in a disruption in the proliferation of neurons destined to become GABAergic interneurons.

To explore whether the behavioral abnormalities observed in stressed animals are related to epigenetic mechanisms, Zhang *et al.*[50] studied the effect of maternal behavior on GABA circuitry in the hippocampus of adult male Long–Evans rats. High maternal licking and grooming behavior of pups correlated with a decrease in DNMT1, a decrease in *GAD1* promoter methylation, and an increase in histone 3 lysine 9 acetylation (H3K9ac) at the *GAD1* promoter followed by an upregulation in the amount of GAD1 mRNA.[50] In another study of adult male BL6/C57 mice,[56] chronic social defeat stress led to the persistent downregulation of BDNF III and IV mRNA in the hippocampus that correlated with an increase in H3K27me2 at the *BDNF* promoter and an increase in avoidance behavior. Roth *et al.*[57] reported that chronic stress in adult male Sprague–Dawley rats results in PTSD-like behavioral changes that are associated with an increase in hippocampal (CA1-DG) *BDNF* promoter methylation and a decrease in BDNF exon IV mRNA. As already discussed, in a study of the adult monkey cortex, early-life maternal deprivation was reported to be associated with changes in hydroxymethylation in the promoters of genes known to be related to neurological or psychiatric disorders.[36]

Thus, there is mounting evidence that early-life stress results in a marked reduction in the potency of GABAergic inhibitory neurotransmission that has broad implications for information processing in the brain. In the cortex and hippocampus, GABA is released from fast-spiking GABAergic presynaptic terminals that impinge on postsynaptic $GABA_A$ receptors located on dendrites, somata, or initial axon segments of glutamatergic pyramidal neurons.[8,58,59] The release of GABA is efficient at synchronizing pyramidal neurons and monoaminergic neuron firing rates and is likely crucial for optimizing cognitive and emotional function. It is therefore plausible that GABAergic neurotransmitter deficits measured in the postmortem brains of SZ patients and in the brains of PRS mice lead to a disruption in intermittent synchronization patterns of pyramidal neuron firing, thereby inducing cognitive and emotional impairment.[58,59]

3. THE EPIGENETIC MODIFICATIONS OF GABAergic AND GLUTAMATERGIC GENES INDUCED BY PRENATAL STRESS IN MICE ARE ALSO DETECTED IN SZ AND BP DISORDER PATIENTS

In a recent study, we tested whether in addition to the impaired migration of inhibitory neuronal progenitors from the median eminence to the cortex,[53–55] the decreased expression of GABAergic genes in the cortex and hippocampus of adult prenatally stressed mice (PRS mice) is also associated with changes in the methylation/demethylation processes operative at these promoters (Erbo Dong *et al.*, unpublished data).[38,39] We measured the expression of DNMT and TET in the FC and hippocampus of adult PRS mice because both enzymes represent epigenetic biomarkers that are increased in the postmortem brains of psychotic patients and because the altered expression of these enzymes predicts a dysfunction of DNA methylation and demethylation that impacts transcription of specific SZ candidate genes including *GAD1*, *RELN*, and *BDNF*.[2] Our results show that the expression of DNMT1 and TET1 is elevated in the FC and hippocampus of adult PRS offspring (like in the brain of SZ patients) and that this elevation is associated with increased binding of DNMT1 and TET1 to specific regulatory domains of the corresponding promoters along with elevated levels of 5MC and 5HMC. Consequently, the expression of critical target genes is downregulated in SZ. Furthermore, the levels of DNMT1 and TET1 mRNA are much higher in PRS mice than the levels of the same enzymes in control mice whether measured at birth or at postnatal days 7, 14, 21, or 60–75. Hence, the offspring of mothers subjected to restraint stress during pregnancy are vulnerable to the development of behavioral abnormalities similar to those observed in psychotic patients including deficits in social interaction, prepulse inhibition of startle (PPI), and fear conditioning (Table 4.1).

4. PRS MICE ARE A PROMISING MODEL FOR STUDIES OF THE NATURAL COURSE OF SZ AND BP DISORDERS

SZ and BP disorders have a natural course, starting with a prodromal phase, a first episode during adolescence or early adulthood with repeated episodes, eventually leading to a deterioration that ensues over subsequent adult years. Hence, the epigenetic history of such complex neurodevelopmental disorders changing at each stage of the illness cannot be adequately studied only in the postmortem brains of chronic SZ patients. In other words, each postmortem sample provides a unique window into a single time point of a chronic illness that is also progressive. To overcome these limitations, we have focused on studying the epigenetic signature and SZ-like behaviors in offspring of PRS mice as a function of postnatal maturation.

The brains of adult PRS mice and postmortem brains of SZ patients have increased levels of DNMT1 and TET1 in common (Table 4.1). PRS mice show higher levels of DNMT and TET in the cortex and hippocampus from birth to adulthood than control mice. Hence, we can infer that the increase of DNMT and TET is likely to be the result of stress-related changes that occur during embryonic life. The importance of stress during embryonic life is indicated by the finding that DNMT and TET are elevated in PRS mice even at birth before differences in maternal care could be a major factor. Further, it has been reported in Swiss albino mice that stressed mothers raising their own pups exhibit maternal care comparable to nonstressed mothers raising their pups.[60]

We cannot assume at the present time whether the time course of changes in DNMT and TET observed in the brains of PRS mice also occurs in the human brain, but it is conceivable that similar neurodevelopmental changes may occur in response to stressful stimulation either *in utero* or during early postnatal life and that such changes may prevent the normal decrease in DNMT and TET expression that progressively occurs from birth to adulthood. This hypothesis is supported by reports that the exposure of pregnant women to psychological stress, malnutrition, or viral infection during pregnancy is associated with an increased incidence of psychosis in their children later in life.[37,44–49]

It is plausible to theorize that in both human and mouse brains, DNMT- and TET-induced GABAergic and glutamatergic stress-related changes during embryonic life are the basis for a disturbance in the reciprocal interactions between GABAergic, glutamatergic, and monoaminergic neurons that is the likely source of the cognitive and emotional disruptions underlying psychotic symptoms. This theory is supported by the fact that psychotic symptoms can be exacerbated by the administration of NMDA receptor antagonists to SZ and BP disorder patients[61] and to PRS mice.[38] To further investigate the hypothesis that prenatal stress, by increasing promoter methylation of GABAergic genes, may be responsible for the epigenetic alterations of GABA–glutamate neuron interactions in PRS mice, we administered valproic acid (VPA) and clozapine to adult PRS mice in doses that are known to act on chromatin remodeling involving *RELN* and *GAD1* promoter demethylation.[62,63] We observed that the combined VPA and clozapine treatment regimen abolished the hyperactivity, stereotypy, and deficits in social interaction and PPI displayed by PRS mice, whereas no effects were observed in vehicle-treated mice. The specificity of drug action in PRS mice is consistent with an epigenetic mechanism underlying PRS behavioral pathology. These preclinical studies support the concept that the PRS model has construct face validity and pharmacological utility as an experimental epigenetic model of SZ/BP disorder. Furthermore, the PRS model theoretically has the potential for use in predicting the course of SZ-like behavioral pathology and more importantly for predicting treatment response at different stages of the illness with particular attention to early detection of the disease.

ACKNOWLEDGMENT

This work is supported in part by RO1MH093348 and RO1MH101049 to A. G.

REFERENCES

1. Ptak C, Petronis A. Epigenetics and complex disease: from etiology to new therapeutics. *Annu Rev Pharmacol Toxicol*. 2008;48:257–276.
2. Grayson DR, Guidotti A. The dynamics of DNA methylation in schizophrenia and related psychiatric disorders. *Neuropsychopharmacology*. 2013;38:138–166.
3. Akbarian S, Kim JJ, Potkin SG, et al. Gene expression for glutamic acid decarboxylase is reduced without loss of neurons in prefrontal cortex of schizophrenics. *Arch Gen Psychiatry*. 1995;52:258–266.
4. Guidotti A, Auta J, Davis JM, et al. Decrease in reelin and glutamic acid decarboxylase$_{67}$ (GAD_{67}) expression in schizophrenia and bipolar disorder: a postmortem brain study. *Arch Gen Psychiatry*. 2000;57:1061–1069.
5. Benes FM, Lim B, Matzilevich D, Walsh JP, Subburaju S, Minns M. Regulation of the GABA cell phenotype in hippocampus of schizophrenics and bipolars. *Proc Natl Acad Sci USA*. 2007;104:10164–10169.
6. Fatemi SH, Earle JA, McMenomy T. Reduction in reelin immunoreactivity in hippocampus of subjects with schizophrenia, bipolar disorder and major depression. *Mol Psychiatry*. 2000;5:654–665.
7. Impagnatiello F, Guidotti A, Pesold C, et al. A decrease of reelin expression as a putative vulnerability factor in schizophrenia. *Proc Natl Acad Sci USA*. 1998;95:15718–15723.
8. Lewis DA, Hashimoto T, Volk DW. Cortical inhibitory neurons and schizophrenia. *Nat Rev Neurosci*. 2005;6:312–324.
9. Wong J, Hyde TM, Cassano HL, Deep-Soboslay A, Kleinman JE, Weickert CS. Promoter specific alterations of brain-derived neurotrophic factor mRNA in schizophrenia. *Neuroscience*. 2010;169:1071–1084.
10. Weickert CS, Hyde TM, Lipska BK, Herman MM, Weinberger DR, Kleinman JE. Reduced brain-derived neurotrophic factor in prefrontal cortex of patients with schizophrenia. *Mol Psychiatry*. 2003;8:592–610.
11. Ikegame T, Bundo M, Sunaga F, et al. DNA methylation analysis of BDNF gene promoters in peripheral blood cells of schizophrenia patients. *Neurosci Res*. 2013;77:208–214.
12. Mill J, Tang T, Kaminsky Z, et al. Epigenomic profiling reveals DNA-methylation changes associated with major psychosis. *Am J Hum Genet*. 2008;82:696–711.
13. Veldic M, Caruncho JH, Liu WS, et al. DNA methyltransferase-1 (DNMT1) is selectively overexpressed in telencephalic GABAergic interneurons of schizophrenia brains. *Proc Natl Acad Sci USA*. 2004;101:348–353.

14. Veldic M, Guidotti A, Maloku E, Davis JM, Costa E. In psychosis, cortical interneurons overexpress DNA-methyltransferase 1. *Proc Natl Acad Sci USA*. 2005;102:2152–2157.
15. Veldic M, Kadriu B, Maloku E, et al. Epigenetic mechanisms expressed in basal ganglia GABAergic neurons differentiate schizophrenia from bipolar disorder. *Schizophr Res*. 2007;91:51–61.
16. Ruzicka WB, Zhubi A, Veldic M, Grayson DR, Costa E, Guidotti A. Selective epigenetic alteration of layer I GABAergic neurons isolated from prefrontal cortex of schizophrenia patients using laser-assisted microdissection. *Mol Psychiatry*. 2007;12:385–397.
17. Zhubi A, Veldic M, Puri NV, et al. An upregulation of DNA-methyltransferase 1 and 3a expressed in telencephalic GABAergic neurons of schizophrenia patients is also detected in peripheral blood lymphocytes. *Schizophr Res*. 2009;111:115–122.
18. Kadriu B, Guidotti A, Chen Y, Grayson DR. The DNA methyltransferases1 (DNMT1) and 3a (DNMT3a) co-localize with GAD67-positive neurons in the GAD67-GFP mouse brain. *J Comp Neurol*. 2012;9:1951–1964.
19. Guidotti A, Ruzicka W, Grayson DR, et al. S-adenosyl methionine and DNA methyltransferase-1 mRNA overexpression in psychosis. *Neuroreport*. 2007;18:57–60.
20. Grayson DR, Jia X, Chen Y, et al. Reelin promoter hypermethylation in schizophrenia. *Proc Natl Acad Sci USA*. 2005;102:9341–9346.
21. Abdolmaleky HM, Cheng KH, Russo A, et al. Hypermethylation of the reelin (RELN) promoter in the brain of schizophrenic patients: a preliminary report. *Am J Med Genet B Neuropsychiatr Genet*. 2005;134B:60–66.
22. Tochigi M, Iwamoto K, Bundo M, et al. Methylation status of the reelin promoter region in the brain of schizophrenic patients. *Biol Psychiatry*. 2008;63:530–533.
23. Gavin DEP, Sharma RP, Chase KA, Matrisciano F, Dong E, Guidotti A. Growth arrest and DNA-damage-inducible, beta (GADD45b)-mediated DNA demethylation in major psychosis. *Neuropsychopharmacology*. 2012;2:531–542.
24. Mitchell CP, Chen Y, Kundakovic M, Costa E, Grayson DR. Histone deacetylase inhibitors decrease reelin promoter methylation in vitro. *J Neurochem*. 2005;93:483–492.
25. Noh JS, Sharma RP, Veldic M, et al. DNA methyltransferase 1 regulates reelin mRNA expression in mouse primary cortical cultures. *Proc Natl Acad Sci USA*. 2005;102:1749–1754.
26. Chen Y, Kundakovic M, Agis-Balboa RC, Pinna G, Grayson DR. Induction of the reelin promoter by retinoic acid is mediated by Sp1. *J Neurochem*. 2007;183:650–665.
27. Kundakovic M, Chen Y, Guidotti A, Grayson DR. The reelin and GAD67 promoters are activated by epigenetic drugs that facilitate the disruption of local repressor complexes. *Mol Pharmacol*. 2009;75:342–354.
28. Dong E, Gavin DP, Chen Y, Davis J. Upregulation of TET1 and down-regulation of APOBEC3A and APOBEC3C in the parietal cortex of psychotic patients. *Transl Psychiatry*. 2012;2:e159.
29. Houston I, Peter CJ, Mitchell A, Straubhaar J, Rogaev E, Akbarian S. Epigenetics in the human brain. *Neuropsychopharmacology*. 2013;38:183–197.
30. Guidotti A, Auta J, Chen Y, et al. Epigenetic GABAergic targets in schizophrenia and bipolar disorder. *Neuropharmacology*. 2011;60:1007–1016.
31. Wyatt RJ, Termini BA, Davis J. Biochemical and sleep studies of schizophrenia. A review of the literature 1960–1970. *Schizophr Bull*. 1971;4:10–44.
32. Tremolizzo L, Doueiri M-S, Dong E, et al. Valproate corrects the schizophrenia-like epigenetic behavioral modifications induced by methionine in mice. *Biol Psychiatry*. 2005;57:500–509.
33. Dong E, Guidotti A, Grayson DR, Costa E. Histone hyperacetylation induces demethylation of reelin and 67-kDa glutamic acid decarboxylase promoters. *Proc Natl Acad Sci USA*. 2007;104:4676–4681.

34. Costa E, Chen Y, Davis J, et al. Reelin and schizophrenia: a disease at the interface of the genome and the epigenome. *Mol Interv*. 2002;2:47–57.
35. Guo JU, Su Y, Zhong C, Ming GL, Song H. Hydroxylation of 5-methylcytosine by TET1 promotes active DNA demethylation in the adult brain. *Cell*. 2011;145:423–434.
36. Massart R, Suderman M, Provencal N, et al. Hydroxymethylation and DNA methylation profiles in the prefrontal cortex of the non-human primate rhesus macaque and the impact of maternal deprivation on hydroxymethylation. *Neuroscience*. 2014;268: 139–148.
37. Koenig JI, Elmer GI, Shepard PD, et al. Prenatal exposure to a repeated variable stress paradigm elicits behavioral and neuroendocrinological changes in the adult offspring: potential relevance to schizophrenia. *Behav Brain Res*. 2005;156:251–261.
38. Matrisciano F, Tueting P, Dalal I, et al. Epigenetic modifications of GABAergic interneurons are associated with the schizophrenia-like phenotype induced by prenatal stress in mice. *Neuropharmacology*. 2013;68:184–194.
39. Matrisciano F, Tueting P, Maccari S, Nicoletti F, Guidotti A. Pharmacological activation of group-II metabotropic glutamate receptors corrects a schizophrenia-like phenotype induced by prenatal stress in mice. *Neuropsychopharmacology*. 2012;37:929–938.
40. Volk DW, Austin MC, Pierri JN, Sampson AR, Lewis DA. Decreased glutamic acid decarboxylase67 messenger RNA expression in a subset of prefrontal cortical gamma-aminobutyric acid neurons in subjects with schizophrenia. *Arch Gen Psychiatry*. 2000;57:237–245.
41. Lipska BK, Lerman DN, Khaing ZZ, Weickert CS, Weinberger DR. Gene expression in dopamine and GABA systems in an animal model of schizophrenia: effects of antipsychotic drugs. *Eur J Neurosci*. 2003;18:391–402.
42. Fatemi SH, Folsom TD, Reutiman TJ, Novak J, Engel RH. Comparative gene expression study of the chronic exposure to clozapine and haloperidol in rat frontal cortex. *Schizophr Res*. 2011;134:211–218.
43. Marco E, Mao CC, Cheney DL, Revuelta A, Costa E. The effects of antipsychotics on the turnover rate of GABA and acetylcholine in rat brain nuclei. *Nature*. 1976;264:363–365.
44. Brown AS. The environment and susceptibility to schizophrenia. *Prog Neurobiol*. 2011;93:23–58.
45. Brown AS, Patterson PH. Maternal infection and schizophrenia: implications for prevention. *Schizophr Bull*. 2011;37:284–290.
46. Mednick SA, Huttunen MO, Machón RA. Prenatal influenza infections and adult schizophrenia. *Schizophr Bull*. 1994;20:263–267.
47. Izumoto Y, Inoue S, Yasuda N. Schizophrenia and the influenza epidemics of 1957 in Japan. *Biol Psychiatry*. 1999;46:119–124.
48. Susser E, Neugebauer R, Hoek HW, et al. Schizophrenia after prenatal famine. Further evidence. *Arch Gen Psychiatry*. 1996;53:25–31.
49. Szyf M. Epigenetics, DNA methylation, and chromatin modifying drugs. *Annu Rev Pharmacol Toxicol*. 2009;49:243–263.
50. Zhang TY, Labonté B, Wen XL, Turecki G, Meaney MJ. Epigenetic mechanisms for the early environmental regulation of hippocampal glucocorticoid receptor gene expression in rodents and humans. *Neuropsychopharmacology*. 2013;38:111–123.
51. McGowan PO, Sasaki A, D'Alessio AC, et al. Epigenetic regulation of the glucocorticoid receptor in human brain associates with childhood abuse. *Nat Neurosci*. 2009;12:342–348.
52. Schmidt MJ, Mirnics K. Neurodevelopment, GABA system dysfunction, and schizophrenia. *Neuropsychopharmacology*. 2014, http://dx.doi.org/10.1038/npp.2014.95.
53. Fine R, Zhang J, Stevens HE. Prenatal stress and inhibitory neuron systems: implications for neuropsychiatric disorders. *Mol Psychiatry*. 2014;19:641–651.

54. Stevens HE, Su T, Yanagawa Y, Vaccarino FM. Prenatal stress delays inhibitory neuron progenitor migration in the developing neocortex. *Psychoneuroendocrinology*. 2013;38:509–521.
55. Uchida T, Furukawa T, Iwata S, Yanagawa Y, Fukuda A. Selective loss of parvalbumin-positive GABAergic interneurons in the cerebral cortex of maternally stressed Gad1-heterozygous mouse offspring. *Transl Psychiatry*. 2014;4:e371.
56. Tsankova NM, Berton O, Renthal W, Kumar A, Neve RL, Nestler EJ. Sustained hippocampal chromatin regulation in a mouse model of depression and antidepressant action. *Nat Neurosci*. 2006;9:519–525.
57. Roth TL, Zoladz PR, Sweatt JD, Diamond DM. Epigenetic modification of hippocampal Bdnf DNA in adult rats in an animal model of post-traumatic stress disorder. *J Psychiatr Res*. 2011;45:919–926.
58. Guidotti A, Auta J, Davis JM, et al. GABAergic dysfunction in schizophrenia: new treatment strategies on the horizon. *Psychopharmacology (Berl)*. 2005;180:191–205.
59. Lewis DA, González-Burgos G. Neuroplasticity of neocortical circuits in schizophrenia. *Neuropsychopharmacology*. 2008;33:141–165.
60. Meek LR, Dittel PL, Sheehan MC, Chan JY, Kjolhaug SR. Effects of stress during pregnancy on maternal behavior in mice. *Physiol Behav*. 2001;72:473–479.
61. Lisman JE, Coyle JT, Green RW, et al. Circuit-based framework for understanding neurotransmitter and risk gene interactions in schizophrenia. *Trends Neurosci*. 2008;31:234–242.
62. Dong E, Chen Y, Gavin DP, Grayson DR, Guidotti A. Valproate induces DNA demethylation in nuclear extracts from adult mouse brain. *Epigenetics*. 2010;5:730–735.
63. Guidotti A, Dong E, Kundakovic M, Satta R, Grayson DR, Costa E. Characterization of the action of antipsychotic subtypes on valproate-induced chromatin remodeling. *Trends Pharmacol Sci*. 2009;30:55–60.

CHAPTER FIVE

Epigenetics of Depression

Sermsak Lolak*, Pim Suwannarat†, Robert H. Lipsky‡,§

*Department of Psychiatry, The George Washington University School of Medicine and Health Sciences, Washington, District of Columbia, USA
†Department of Genetics, Mid-Atlantic Permanente Medical Group, Rockville, Maryland, USA
‡Inova Neurosciences Institute, Inova Health System, Falls Church, Virginia, USA
§Department of Molecular Neuroscience, The Krasnow Institute for Advanced Study, George Mason University, Fairfax, Virginia, USA

Contents

Abstract

Major depressive disorder (MDD) is a leading cause of disability worldwide and is associated with poor psychological, medical, and socioeconomic outcomes. Although much has been learned about the etiology and treatment options of MDD over the past decade, there remain unanswered questions that pose challenges to improving acute and chronic outcomes for those with MDD. MDD is a clinically heterogeneous disorder. Genetic studies to date have indicated a number of genes, including transporters, neurotransmitters, neurotrophins, and their associated signaling networks that may predispose individuals to MDD and may also predict treatment outcomes. However,

Progress in Molecular Biology and Translational Science, Volume 128
ISSN 1877-1173
http://dx.doi.org/10.1016/B978-0-12-800977-2.00005-X

twin studies indicate that genes account for only a small degree of the variation in MDD. Thus, other mechanisms, through epigenetic marks, may act to form a molecular memory of previous gene-to-environment interactions and to establish vulnerabilities (or, conversely, resistance) to MDD. Current evidence supports a role for pre-, peri-, and early postnatal adversities and stressful life events into adulthood affecting epigenetic patterns, providing a mechanistic foundation to develop epigenetic marks as biomarkers for MDD. This review presents the evidence supporting a role for epigenetic effects in MDD and in treatment response. We also discuss the controversy behind modulating epigenetic mechanisms in long-term antidepressant pharmacotherapy.

1. INTRODUCTION

1.1. Diagnosis of depression

Major depressive disorder (MDD), or major depression, is an established psychiatric disorder characterized by a combination of clinical symptoms in affective–cognitive (depressed mood, hopelessness, worthlessness, excessive guilt, or suicidal ideation), neurovegetative (poor sleep, appetite, energy, and concentration), and behavioral (loss of interest and inability to function) domains.

According to *the Diagnostic and Statistical Manual of Mental Disorders* (5th edition) (DSM-5),[1] MDD is defined by one or more major depressive episodes (MDEs) and the lifetime absence of mania and hypomania. An MDE is defined by five or more symptoms out of nine over at least a 2-week period that causes significant impairment or distress. One of these symptoms must be depressed mood or anhedonia (loss of interest or pleasure). In addition, the episode is not attributable to a substance or medical condition nor is it better explained by a psychotic disorder such as schizophrenia or schizoaffective disorder. A significant and controversial change in the criteria of MDD in DSM-5 that is different from DSM-IV is that the "bereavement exclusion" for the diagnosis of MDE has been removed, and the note calling for clinical judgment when diagnosing MDD in the context of a significant loss has been added. This change will potentially have a significant impact on the epidemiology, clinical care, and research of MDD in the direction that will be unclear. For example, while the removal of bereavement exclusion may be advantageous for the study of environmental factors in the etiology of MDD as bereavement could be the potential confounder, the replaced "note" in DSM-5 has a potential to blur the line between pathology and normality in an unpredictable way, which can affect diagnosis reliability and epidemiological data of the disorder.[2,3] Only time and further studies will tell how much these changes are impacting the field.

1.2. Demographics of depression

Depression is a common illness worldwide, with an estimated 350 million people affected (http://www.who.int/mediacentre/factsheets/fs369/en/, accessed December, 2013). In 2010, mental and substance use disorders were the leading cause of years lived with disability (YLD) worldwide (175.3 million or 22.9% of all YLD).[4] According to the World Health Organization (WHO), MDD was the fourth leading cause of disability worldwide. It is projected that, by 2020, depression will be the second leading cause of disability throughout the world, trailing only ischemic heart disease.[5] The prevalence of depression varies across countries, but in general, lifetime prevalence is estimated to be higher in high-income countries.[6] In the United States, the lifetime prevalence estimate of MDD is 16.5%.[7] A recent WHO survey from 18 countries showed 12-month and lifetime prevalences of MDE to be 14.6% and 5.5% in high-income countries compared with 11.1% and 5.9% in low- and middle-income countries.[8]

The impact of depression on quality of life, economic productivity, and financial burden has been extensively discussed and is beyond the scope of this review.

1.2.1 Effect of gender and age on depression

Women in general have a higher prevalence of depression compared with the age-match men. This may be attributable to the difference in the brain's neurochemical response to serotonin, among other neuronal functions.[9] Evidence suggests a gonadal hormone influence on complex interactions between serotonin and neural circuits that mediate the stress axis, which could explain some of the sex differences in the response of antidepressant (AD).[10]

There is an increased incidence of depression in males with decreased testosterone level, either age-related or illness-related, for which androgen therapy may be effective.[11] Similarly, change in female sex hormones is associated with the development of depression.[12] Animal models examining the interaction between stress pathways and hormonal effects and effects of early-life stress provided some information on the sex differences in psychopathology, in addition to the true gene differences encoded in sex chromosomes.[13]

1.2.2 Depression as a comorbid disorder

MDD is also associated with elevated risk of onset, persistence, and severity of a wide range of chronic physical disorders and with increased early mortality due to an even wider range of physical disorders beyond suicide. MDD is a consistent predictor of the subsequent first onset of many medical diseases including coronary artery disease, stroke, diabetes, and certain types of cancer.[14]

Evidence suggests that MDD is a risk factor for cardiovascular disease and is associated with the increased mortality risk in patients with established heart disease.[15] A number of plausible mechanisms linking depression and poor physical outcomes have been suggested. This includes both biological mechanisms, such as dysregulation of hypothalamic–pituitary–adrenal (HPA) axis, and poor health behaviors, such as smoking and decreased adherence to treatment.[16] A discussion of the underlying biochemical, genetic, and epigenetic mechanisms follows.

1.3. The etiology of depression

The etiology of depression is still not well understood. There is simply no "unifying theory" that can explain all clinical depression, likely because of the heterogeneity of the depressive syndromes and diagnostic criteria that rely heavily on the subjective quality and quantity of symptoms. The popular but overly simplistic "chemical imbalance" theory of depression, particularly in monoamine (serotonin, noradrenaline, and dopamine) neurotransmission deficits, has failed to produce reliable evidence as a causal explanation of depressive syndrome. However, multiple animal and clinical studies support the role of serotonergic defect in MDD. Consistent with previous reports on the role of tryptophan, a precursor of serotonin, in depressed individuals, MDD patients whose depression is in remission have decreased serotonin levels in the brain and often experience a brief relapse of symptoms when they are placed on a diet deficient in tryptophan (tryptophan depletion).[17] However, the clinical evidence that current ADs whose mechanism of action increases monoamine neurotransmission are efficacious does not necessarily mean that the opposite is true for the pathophysiology of depression.[18] Unfortunately, newer ADs such as selective serotonin reuptake inhibitors (SSRIs) or serotonin and norepinephrine reuptake inhibitors (SNRIs) failed to improve efficacy. Only one-third of the patients respond to SSRIs, such as citalopram, following initial treatment.[19] In addition, the recent findings that intravenous ketamine, which exerts its effect on NMDA glutaminergic systems, can rapidly improve symptoms in treatment-resistant depression[20] prompts further exploration of underlying neurobiological mechanisms of depression. Interestingly, ketamine response may be genetically related as some of the nonresponders are carriers of the Met allele of Val66Met (rs6265), a single-nucleotide polymorphism (SNP) that is associated with attenuated brain-derived neurotrophic factor (BDNF) functioning.[21,22] In addition, as we shall see, epigenetic mechanisms play a major role in regulating BDNF expression.

1.3.1 The neurotrophin hypothesis of MDD

BDNF, a neuroprotective protein that regulates neuronal survival, growth, and differentiation, has been implicated in the pathogenesis of depression and a potential therapeutic mechanism of ADs.[23] The interaction of serotonin and neurotrophic systems in MDD is supported by the fact that during tryptophan depletion, BDNF levels are increased in healthy volunteers but in remitted MDD patients, the compensatory response is not seen.[24] BDNF synthesis and release are targeted by activation of glutamate receptors. Stress-induced dysregulation of BDNF-ERK1/2-CREB-Bcl2 cascade may produce atrophy of vulnerable neurons and dendrites, perhaps associated with hippocampal atrophy seen in MDD.[25] This forms the basis of the "neurotrophin hypothesis" of MDD.

Direct infusion of BDNF to the hippocampus produces AD effect, while the effect is opposite for ventral tegmental area.[26] Depression is associated with reductions in serum, hippocampal, and prefrontal cortex BDNF; however, these reductions are reversed with AD treatment.[27,28] In addition, injection of interferon-alpha, a medication well known to provoke depressive symptoms, is associated with reduced BDNF and hippocampal neurogenesis.[26,29] It has been suggested that most known effective AD treatments including SSRI, tricyclic AD, electroconvulsive therapy, and deep brain stimulation positively influence hippocampal BDNF expression and neurogenesis.[26] In addition, the AD responses are lost when the *BDNF* gene is not functional.[30] Association studies of the *BDNF* gene and depression have produced mixed reports,[25] although in a recent pilot study of elderly patients with depression, an improvement in depressive symptoms measured by the Geriatric Depression Scale after treatment with escitalopram for 2 months was significantly associated with increased serum BDNF levels.[31]

Stress leads to a reduction in BDNF expression, suggesting an impairment in neuroplasticity, which may contribute to the development of depression.[32] However, it may be too simplistic to assume a direct causal link of BDNF to depression and AD response as these effects are regional and AD-specific in the context of other genetics and environmental backgrounds.[33] Given that there are also negative studies, clearly, there is a need for further research to elucidate a mechanism through which BDNF modulates depressive symptoms and treatment.[26,34]

1.3.2 Role of the HPA axis in depression

Dysregulation of HPA axis function secondary to stress is associated with the development of depression. Stressful stimuli, which are associated with the development of MDD, have a physiological underpinning based on increased HPA axis activity, a glucocorticoid receptor (GR)-dependent mechanism.[35] The GR is a central nuclear receptor that regulates cellular responses to glucocorticoids like the stress hormone cortisol. GR mediates the negative feedback regulation needed to terminate the response initiated by stress. Acute stress can increase cognitive performance following the stressful event in both animal models and humans.[36,37] However, hyperactivity of the HPA axis in MDD, which has become one of the most reliably reported neurobiological characteristics associated with affective disorders, carries a greater risk of negative consequences. Hyperactivity of the HPA axis is thought to occur because of reduced negative feedback control by cortisol (corticosterone in rodent models). Altered negative feedback regulation is thought to be an underlying mechanism in the pathophysiology of MDD. The dexamethasone suppression test (DST) has been adapted to examine nonsuppression of cortisol in affective disorders and has been suggested as a biomarker for stress response.[38,39] The DST determines adrenal gland function by measuring cortisol levels in the blood, saliva, or urine in response to exogenous administration of dexamethasone, as synthetic glucocorticoid hormone. A low dose, typically 1–2 mg, causes a decrease in cortisol levels. Originally used as a diagnostic test for Cushing's syndrome, nonsuppression of cortisol by dexamethasone provides a measure of negative feedback inhibition of the HPA axis. In addition, adrenocorticotropic hormone (ACTH) levels have also been used to determine robustness of the negative feedback pathway. During normal HPA axis activity, exogenous dexamethasone inhibits release of corticotropin-releasing hormone and release of ACTH, and both effects result in a reduction of cortisol release from the adrenal gland. There has been considerable evidence supporting the idea that resistance to dexamethasone challenge is associated with MDD.[40,41] Whether these alterations in HPA axis regulation are limited to the acute stage of MDD or whether they persist even through recovery remains unclear. Recent findings by Behnken *et al.*[42] support the idea that in recovered MDD patients, these individuals frequently have poorer cognitive performance relative to healthy control subjects. This reduced performance correlated with increased cortisol response to DST. Complicating interpretation of DST responses is the observation that increasing age also reduces negative feedback sensitivity.[43,44]

While most studies involving HPA axis sensitivity have focused on peripheral responses (monocyte GR levels in response to challenge and saliva and plasma cortisol levels), changes in the brain at the anatomical, cellular, and molecular levels have been observed in postmortem samples and during neuroimaging of depressed patients, which affects numerous brain regions, including the prefrontal cortex, hippocampus (cognition and memory), amygdala (emotion), and nucleus accumbens (reward processing and HPA axis). Decreased hippocampal volume has been seen in depressed individuals,[45,46] along with increased metabolism by the amygdala.[47] Decreased activation of the nucleus accumbens in response to rewarding stimuli and increased HPA axis activity has also been observed. Because HPA axis effects converge on the GR, it is likely that that impairment of HPA axis activity seen in MDD is likely to operate at the level of genome, initiating cellular processes that inhibit both neurogenesis and neuroprotection, events associated with cellular loss and adverse changes in learning and memory.[48] This is supported by findings from animal models of depression and clinical studies showing that excessive glucocorticoids in the brain, particularly the hippocampus, are the basis of glucocorticoid-mediated neurotoxicity.[49] One clinical model is Cushing's syndrome, typically associated with the use of glucocorticoid medications. Interestingly, about 60% of individuals with Cushing's syndrome (also known as hypercorticism) have depressive symptoms that respond when cortisol levels are restored to normal.[50] Cushing's syndrome is etiologically different from Cushing's disease, in that the pituitary adenoma produces high levels of ACTH that creates the aberrant increase in cortisol in those tumor patients. Cushing's disease patients have higher scores on subscales for depression compared to age-, gender-, and educational level-matched controls.[51]

It is notable that chronic exposure to an AD is associated with normalization of HPA function, which can precede improvement of behavioral symptoms of clinical depression.[52] Normalization of HPA response from chronic AD exposure is associated with hippocampal neurogenesis, suggesting the role of neuroplasticity in the etiology and treatment of depression.[53]

1.3.3 Animal models of stress and its role in depression

Animal models, as well as neuroimaging and postmortem studies on the effects of chronic stress, which can precipitate or exacerbate depression, provide valuable information. The development of animal stress paradigms especially in rodents, along with knowledge on underlying neurocircuitry, has helped advance understanding of the pathophysiology and treatment of MDD.[28] Chronic animal exposure to stressful stimuli, such as prolonged physical stress or social subordination, can produce anhedonia-like symptoms. This effect is rarely seen following exposure to acute stress.

Male tree shrews (*Tupaia belangeri*) provide a promising model of the chronic psychosocial stress paradigm. Tree shrews and humans have similar stress response-related proteins, including corticotropin-releasing factor, glucocorticoid, and mineralocorticoid receptors. Pairs of male tree shrews in visual and olfactory contact develop a stable dominant/subordinate relationship. The subordinates show behavioral, neuroendocrine, and central nervous activity changes, similar to the signs and symptoms during episodes of depression in patients. In addition, these changes have predictive validity as these symptoms can be treated with AD medications but not anxiolytic agents.[54] Male tree shrews under chronic psychosocial stress have significantly decreased hippocampal volume and proliferation rate of the granule precursor cells in the dentate gyrus. These changes can be prevented by treating the animals with AD medication such as clomipramine and tianeptine, illustrating the link between depressive symptoms and alterations in brain metabolism, hippocampal volume, and neurogenesis.[55]

1.3.4 MDD is associated with structural changes in the brain

MDD is associated with structural and functional changes in several brain regions including the prefrontal cortex, hippocampus, amygdala, and ventral striatum.[18,26] This occurs as a consequence of modulation and alteration of neurotransmission and neurotrophic factors. The nucleus accumbens, a reward processing center, is also affected and thought to underlie the anhedonia symptoms of MDD.[28] The hippocampal formation contains receptors associated with the regulation of stress response and thus is particularly sensitive to the effects of stress and depression. Hippocampal abnormalities may explain the clinical syndrome of decreased attention, concentration, explicit memory, and motivation typically seen in depressed patients. In addition to these structural and functional changes, neural plasticity, the ability of neurons and neural elements to adapt in response to intrinsic and extrinsic signals, has been postulated as a possible mechanism that can potentially bridge different findings.[34] The adult hippocampal neurogenesis theory of depression has been suggested, as there is a consistent evidence of reduced hippocampal size as well as reduced hippocampal neurogenesis in patients with depression that may be the result of cumulative stress.[32]

1.4. AD treatments

AD medications, which primarily act on monoamines such as serotonin, noradrenaline, and dopamine or a combination of these neurotransmitters, have been the mainstay of treatment of depression. According to the American Psychiatric Association treatment guidelines for patients with MDD,[56] AD medications can be used as an initial treatment modality for patients with mild, moderate, or severe MDD. Clinical features that may suggest that medications are the preferred treatment modality include a history of prior positive response to AD medications, the presence of moderate to severe symptoms, significant sleep or appetite disturbances, agitation, patient preference, and anticipation of the need for maintenance therapy. However, the efficacy of ADs has not been impressive. A large naturalistic study finds that only a third of MDD patients achieve complete remission after a single AD trial.[57] In addition, it has been observed that AD treatment generally has a delayed clinical response up to 4–6 weeks, although recent meta-analyses have shown support for earlier onset of improvement within the first week or two of treatment.[58]

Delayed onset of response to AD was previously explained as time required for serotonin autoreceptor to sensitize. However, the validity of the monoamine hypothesis as a sole etiology of depression is questioned by the findings that disrupting portions of the serotonergic system fails to induce a depressive phenotype.[34] It is plausible that the delayed response targeting monoaminergic pathways may be mediated through long-term adaptations, such as epigenetic effects.[59,60] It is notable that certain ADs exert their action through various epigenetic mechanisms by decreasing methylation levels of DNA, influencing microRNAs, and modification of histones.[59]

1.5. AD treatment response

The mechanisms of response to AD therapy are poorly understood. Variations in genes implicated in 5-HT neurotransmission may interact with environmental factors to influence AD response. El Hage and Powell[10,61] reviewed possible mechanisms and main predictors of poor response to AD treatment, which include (1) clinical correlates such as old age, somatic comorbidities, and poor compliance in relation to socioeconomic status; (2) drug metabolism such as alteration in drug metabolism due to genetic differences, concomitant medication, diet, or smoking; (3) blood–brain barrier alteration, especially due to the effect to P-glycoprotein from polymorphisms of genes drug interaction; (4) differences in brain structures, for example, lower alpha activity in posterior regions and lower baseline rostral anterior cingulate cortex activity in AD nonresponders; (5) neurotransmission, such as the polymorphisms of the serotonin or noradrenaline transporter (NET) genes, which may also interact with environmental factor in determining response to AD; (6) neural plasticity, such as alteration of BDNF due to polymorphisms and other factors, alteration in adult hippocampal neurogenesis, and zinc deficiency; and (7) hormonal targets, including defect in HPA axis regulation, polymorphisms of genes encoding GR complex such as FKBP51, and hypothyroidism as evidenced by role of T3 in mediating AD response.

Stress, which is associated with depression, can suppress hippocampal neurogenesis in animal models.[62] There are considerable evidences that AD can prevent and reverse stress-induced impairment of hippocampal neurogenesis and that the behavioral effect of AD is abolished when neurogenesis is prevented.[62] On the other hand, this may not necessarily be the case, as there is also evidence showing that blockade of hippocampal neurogenesis does not reduce AD efficacy in animal models,[63] and so far, there is a paucity of such evidence in humans as well.[64] Thus, the hippocampal neurogenesis theory alone may not be sufficient. Interestingly, chronic, but not acute, treatment with AD, which is comparable to the duration of clinical effects in humans, can reverse the anhedonia in animals exposed to chronic stress.[28]

A considerable evidence suggests that chronic treatment of AD increases expression of BDNF, particularly in the hippocampus and prefrontal cortex.[32] Several classes of ADs (with chronic exposure) and electroconvulsive therapy (ECT) enhance hippocampal neurogenesis. Perera *et al.*[65] demonstrated that repeated electroconvulsive shock (ECS)—the animal analog of ECT—increased precursor cell proliferation in the subgranular zone of the dentate gyrus in adult monkey hippocampus, the delayed onset of which coincides with the delayed clinical response time of ADs.[34]

2. GENETICS OF MAJOR DEPRESSION

Depression aggregates in families. However, psychiatric disorders are not inherited in a Mendelian pattern. Rather, multiple susceptibility genes are associated with psychiatric disorders, each contributing a small effect with variable expression of disease. Twin studies estimate the heritability of depression to be about 37% but can increase up to 70% when severity, relapse rate, and age of onset are considered.[59,66] In spite of this, the search for genetic causes has been unsuccessful to date, the "depression genes" have not yet been identified, and genetics association studies have not yet uncovered consistent genetic risk modifiers.[33] This may be in part due to the fact that genetics alone does not account fully; rather, the complex interplay of genetics, environmental, and other factors that determine the individual's susceptibility and resilience to the disease also plays a role in the etiology of MDD. Consequently, in comparison with other multifactorial medical diseases such as type 2 diabetes or cancer, the progress in understanding of molecular biology of depression has been slow.[18] The etiology of MDD likely involves multiple genes, genes and environment interactions, neurobiological systems, and epigenetic effect.[25,67]

Linkage studies have identified chromosomal regions, genes, and SNPs of possible interest. However, for the most part, these studies have not been replicated. However, this approach has benefits of correlating data to future studies such from candidate gene studies and genome-wide association studies (GWAS).[68]

2.1. Candidate gene association studies

Candidate gene association studies (CGAS) are an alternative approach to discover responsible genes based on an educated guess of pathophysiological mechanisms of MDD. One such mechanism tests the monoamine hypothesis of depression, focusing on serotonin. Serotonin is one of the monoamines associated with the pathophysiology and treatment of MDD. The serotonin transporter (5-HTT, encoded by the *SLC6A4* gene) regulates concentration of synaptic serotonin and is the site of action of several ADs, mainly SSRIs, SNRIs, and tricyclic ADs. A functional polymorphism in the *SLC6A4* gene upstream promoter region (*HTTLPR*) has been extensively studied as a potential site associated with MDD and its association with AD response. However, the results so far are conflicting (reviewed in Refs. 25 and 69).

As originally described, *HTTLPR* was described as an insertion or deletion of a 44-base pair-long region, giving rise to short (S) and long (L) alleles of the promoter region (Fig. 5.1). The "S" allele was associated with reduced *SLC6A4* mRNA expression.[71] Meta-analyses supported small, but statistically significant, associations between the "S" allele and bipolar disorder, suicidal behavior, and depression-related personality traits but not to MDD itself.[72]

2.1.1 Genetic aspects of the serotonin transporter

Changes in expression of *SLC6A4* are driven in part by the promoter polymorphism *HTTLPR* and involves modulation of the HPA axis,[61] thus linking stress responses to serotonin neurotransmission. Transcriptional effects of HTTLPR have been the focus of numerous studies, initially reported by Lesch *et al.*[71] Since then, numerous studies have been performed on populations of MDD patients, including a large cohort from the Sequenced Treatment Alternatives to Relieve Depression (STAR*D).[73–75] Genetic associations of variation at the *SLC6A4* gene with MDD and with AD treatment have been variable, although a recent meta-analysis suggested that reduced *SLC6A4* expression was a risk factor for developing MDD when stressed.[76] We provided support for the idea of functional variation at *HTTLPR* by describing an A > G substitution in the L allele (rs25531) that created a binding site for the AP-2 transcription factor (Fig. 5.1). In lymphoblast cells, which express HTT, the allele combination, LG (designated LG), reduces *SLC6A4* mRNA levels and is nearly equal to that of the S allele. The role of functional variation at the *SLC6A4* gene is of interest due to variable responses by patients on SSRI therapy. We showed that the *HTTLPR* LA allele was associated with reduced adverse effects in adult outpatients with major depression treated with citalopram,[73] but the mechanisms underlying the association are unknown. Therefore, we investigated changes in HTT mRNA levels by RT-PCR and determined

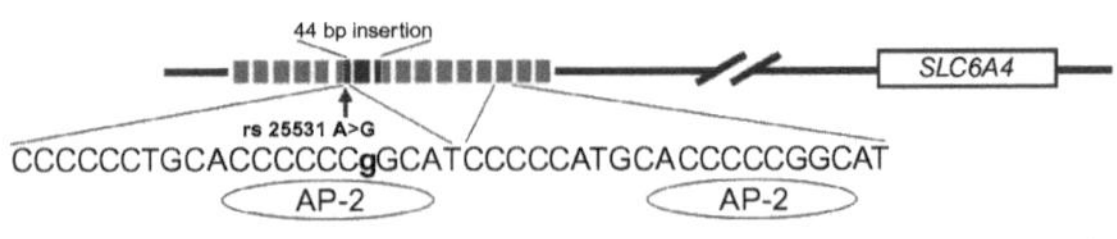

Figure 5.1 Schematic diagram of the SLC6A4 gene promoter region showing the A > G SNP (rs25531) within the polymorphic *HTTLPR* region. Note that the G allele of SNP rs25531 occurs within the sixth unit repeat of the *HTTLPR* and creates a functional AP-2 binding site within the "L" or "long" allele of *HTTLPR*[70] and is shown by an arrow. Note that a second AP-2 site is present in the 14th repeat unit of the *HTTLPR*. Unit repeats are depicted as green-shaded boxes. (See the color plate.)

AP-2 occupancy of the *HTTLPR* region using a chromatin immunoprecipitation (ChIP) protocol with lymphoblast cells of known *HTTLPR* genotypes prepared from drug-free patients with MDD that were isolated and maintained in culture and then treated with citalopram. We found that expression of *SLC6A4* mRNA was stable with increasing culture passage, provided that the cells were growing at the same rate. Citalopram (100 ng/ml) increased *SLC6A4* mRNA levels more than twofold over baseline after 3 h in the cells of the LA/LA genotype. Lymphoblasts of the S/S or LG/LG genotype retained baseline *SLC6A4* mRNA levels following citalopram treatment. ChIP assay results paralleled the *SLC6A4* mRNA findings in that LA/LA lymphoblasts treated with citalopram showed reduced AP-2 occupancy of the *SLC6A4* promoter at a site 3′ of the *HTTLPR* as soon as 30 min after addition of citalopram to the culture medium (Fig. 5.2).[77] S/S and LG/LG lymphoblasts had increased AP-2 occupancy following citalopram treatment. Taken together, these findings provided a mechanism to explain increased *SLC6A4* gene expression in individuals with the LALA genotype and its association with reduced adverse effects in depressed patients undergoing citalopram pharmacotherapy. Since the AP-2 site examined in this study was downstream of the A > G variant, this may represent an epigenetic effect and could potentially provide a molecular mechanism for a drug-induced environmental effect

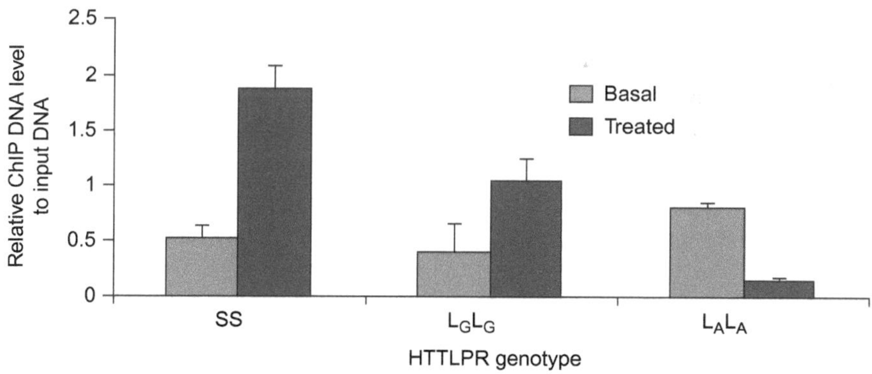

Figure 5.2 An AP-2 transcription factor site in the *SLC6A4* gene promoter. Relative ChIP DNA levels normalized to input of each citalopram-treated sample, before and after 30 min citalopram treatment, are shown by means ± standard error. The results show that AP-2 binding to the 3′ site of *HTTLPR* is decreased significantly in the cell lines with the L_AL_A genotype following citalopram treatment. Since the downstream AP-2 interaction within the promoter region can be observed at a distance from the A > G SNP, citalopram has effects on protein–DNA interactions, which are central to epigenetic mechanisms. (See the color plate.)

interacting with a genetic polymorphism. Long-term effects of SSRI treatment on *SLC6A4* expression by patient-derived lymphocytes will need to be determined before any conclusions on epigenetic mechanisms can be made.

2.2. Genome-wide association studies

GWAS, made possible by advancements in genotyping technology, is a new comprehensive and less biased method to study the genetic component of disease without predetermined hypothesis. However, to reduce false-positive results, correction for multiple testings must be applied.[59] The standard in the field has been to set genome-wide significance at $P \leq 5.0 \times 10^{-8}$.[78] Large meta-analyses in order to obtain combined samples provide better results. Unfortunately, GWAS on MDD has yielded inconsistent results, and most if not all markers do not withstand the correction for multiple testings.[59] However, these studies do support interesting candidate genes and genomic regions for further study.[67] A recent and the largest genome-wide analysis of MDD to date, "mega-analysis," analyzing over 1.2 million autosomal and X chromosome SNPs in 18,759 subjects failed to identify any genetic marker for MDD.[79]

2.3. Genetics of AD treatment response

It is postulated that genetic factors contribute for about 50% of AD response.[69] Pharmacogenetics has promising potential in that it may help clinicians select medications that likely have the best efficacy and the fewest adverse effects for a particular patient based on the patient's genetic markers. Possible genetic contribution to AD response is evident in familial patterns of response. To date, most pharmacogenetic studies have explored genes related to metabolism, neurotransmitter receptors and transporters, and second messenger system.[69] Genetic association studies have identified many polymorphisms suggestive of genetic contribution to AD response. Responsible genetic markers have been identified primarily by two methods, CGAS and GWAS. Singh *et al.*[80] reviewed CGAS of AD efficacy and MDD and found that two studies had positive findings after stringent statistical correction of multiple polymorphism testings. In the first study, Shi *et al.*[81] identified SNPs in the calcium-/calmodulin-dependent protein kinase (CaMK) pathway associated with AD response among Chinese females. No significant interactions between candidate genes and environmental effects were observed. In the second study with positive findings, Singh *et al.*[82] found that a polymorphism of ATP-binding cassette family of transporter proteins, subfamily B (MDR/TAP), member 1 (ABCB1), a key component of the blood–brain barrier, predicts escitalopram dose needed for MDD remission. The dose needed for C carriers at rs1045642 SNP of ABCB1 was twice as high, suggesting implication in clinical practice.

Laje and McMahon[83] reviewed three GWAS of AD response, the STAR*D sample ($n=1953$),[84] the Munich AD Response Signature sample ($n=339$),[85] and the Genome-Based Therapeutic Drugs for Depression sample ($n=706$).[86] Unfortunately, none reported findings that reached genome-wide threshold significance. In addition, the top results from each study do not overlap at the level of genes or alleles. This, however, is unlikely to be due to a lack of genetic effect; rather, it represents a challenge of conducting GWAS because of lack of clear phenotype definitions of depression and AD response, genotyping techniques, and possible epigenetic contribution to AD response and inadequate sample size.[22] In addition, the interaction of genes and environment may have contributed to the results of pharmacogenetic studies. Randomized clinical trials evaluating the utility of genetically guided AD treatment are needed.[80]

Meta-analyses determining the role of HTTLPR in AD response also found inconsistent results.[69,80] Previously, Serretti *et al.*[87] analyzed data from 15 studies and found significant association between two copies of the "S" allele and remission rate and presence of either one or two copies of the "S" allele and AD response rate. However, Taylor *et al.*[88] failed to replicate the results in the meta-analysis of 28 studies. Both meta-analyses observed significant ethnic difference, however. Porcelli *et al.*[89] did a meta-analysis on 33 studies and found that in Caucasians, but not Asians, HTTLPR may be a predictor of AD response and remission.

In other studies, Binder *et al.*[90] suggested that genetic variation at the *FKBP5* gene was associated with outcome of AD treatment and recurrence in major depression. The *FKBP5* gene encodes FK506-binding protein (FKBP5), a chaperone protein important in regulating the HPA axis by decreasing cortisol binding and preventing translocation of the receptor complex to the nucleus.[91] FKBP5 also contains GR activity. In further analyses by Binder and colleagues, it was shown that the TT-homozygote genotype of *FKBP5* SNP rs1360780 was most strongly associated with better treatment response and more previous depressive episodes. No association of *FKBP5* markers with disease status was observed.

In a separate candidate gene study of participants in the STAR*D and healthy controls, we determined in case–control analysis that SNP marker rs1360780 was significantly associated with disease status in a White non-Hispanic sample.[92] This marker is of particular interest since it has been recently suggested that the A allele may form a TATA boxlike element that could increase *FKBP5* transcription.[93] In addition, a significant association was found between the *FKBP5* marker rs4713916 and remission following a course of citalopram monotherapy. Markers rs1360780 and rs4713916 were in a strong linkage disequilibrium in the White non-Hispanic but not in the Black population. There was no significant difference in the number of previous episodes of depression between genotypes at any of the three markers.[92] Taken together, these data support a role for genetic variation of the *FKBP5* in susceptibility to MDD and for AD treatment outcome. The role of epigenetic effects acting on *FKBP5* gene variants, such as rs1360780, in long-term dysregulation of the HPA axis and susceptibility to developing disorders is discussed in the next section.

3. EPIGENETICS OF MAJOR DEPRESSION

3.1. Environmental and developmental effects

Stressful life events, in particular those affecting early development of the central nervous system, are being established as risk factors for the development of MDD. These environmentally based mechanisms have the potential to modify genetically determined predisposition to disease from fetal development to adolescence. This idea is supported in the case of MDD, where heritability of the disorder is modest, accounting for approximately 30–37% of genetic susceptibility based on twin studies.[94] Additionally, GWAS have failed to provide consistently significant findings for MDD as outlined in the previous section. Thus, other factors, or a "missing heritability," are thought to account for paucity of genetic events. One such environmentally based effect is termed a stressor. Stressors may be physiological or psychological. The effect of a stressor may be appraised by individuals differently, leading to responses that neuropsychologist term "resilient" or "susceptible." By example, Bonanno *et al.*[95] presented evidence that individuals having fewer prior traumatic events were more likely to have resilience following the September 11 terrorist attack. Of note in this study was the follow-up period, which was 6 months. What are the long-term effects of stressors? A number of studies, including that of Fossion *et al.*[96] examined resilience and susceptibility to depression and anxiety in older Nazi Holocaust survivors who as children had endured multiple traumatic events. In general, individuals exposed to childhood trauma become less able to cope with trauma(s) as adults. In addition, resilience is apparently diminished by multiple traumas, from childhood as well, and by trauma suffered by adults.[96]

Environmental modifiers of genes impact phenotypes. This hypothesis was tested by Caspi *et al.*[97] for a gene–environment interaction predicting susceptibility to depression, where a genetic variant controlling the expression of the serotonin transporter predicted susceptibility to depression only in the context of early adulthood traumatic experiences. The number of stressful life events was significantly correlated with the probability of developing MDD when there was either one or two copies of the "S" allele of *HTTLPR*.[97] Subsequent meta-analysis has shown conflicting results. Risch *et al.*[98] showed that while the number of stressful life events was significantly associated with depression, there was no association between *HTTLPR* genotype itself and depression nor was there an interaction effect between genotype and stressful life events on depression. On the other hand, another larger meta-analysis published in 2011 showed strong evidence that the studies published to date support the hypothesis that *HTTLPR* moderates the relationship between stress and depression.[76] Although still the subject of scientific debate and refinement, the Caspi *et al.*[97] study has been the conceptual framework for studies designed to understand the molecular basis of MDD.

Epigenetic mechanisms appear to be well suited to provide the molecular basis for genomic changes that alter susceptibility (and resilience) that affect clinical outcomes.

3.2. Effects of early-life stress on epigenetic mechanisms

By the same measure, environmental effects such as early-life stress and its interdependence on epigenetic factors are likely to make a strong contribution to the risk of depression at the gene level. The major epigenetic mechanisms operating on the genome, namely, histone acetylation and methylation, DNA methylation, and microRNA, have been provided in earlier chapters of this book, so it is unnecessary to describe them here. But support for epigenetic mechanisms modifying genes to produce altered phenotypes comes from both animal model and human studies. For example, in an animal model of early-life stress characterized by decreased maternal care, increased DNA methylation at the promoter of the GR gene (*NR3C1* in humans and *Nr3c1* in rodents), which persisted into adulthood, was associated with a perturbed HPA axis function.[99] Increased DNA methylation correlated with reduced *Nr3c1* gene expression. In human studies on identical twins, DNA methylation and histone acetylation patterns across the genome and at specific loci differed among twin pairs with age, supporting genomic changes by environmental influences and demonstrating different developmental outcomes.[100] Using the GR–serotonin association alone, it is clear that the etiology of depression involves molecular, cellular, and neuronal network changes in addition to the monoamine pathway. This conclusion comes from clinical findings: ADs acting to block the reuptake of monoamines or inhibiting their degradation at the synaptic cleft to increase serotonin and noradrenaline transmission require a few weeks of treatment to see a response in depressive symptoms. Furthermore, monoamine depletion in medication-free depressed patients with a family history of depression whose symptoms were in remission led to decreased mood, but decreased mood was not seen in healthy subjects.[101] Therefore, the etiology of depression is complex and involves alterations besides the monoamine system.

3.2.1 Intersection of stress and epigenetics

3.2.1.1 FK506-binding protein 5

As previously noted, the GR is a central receptor and a transcription factor in regulating the HPA axis. In fact, HPA axis dysregulation and GR resistance can be governed by changes in FK506-binding protein 5 (FKBP5) function.[91] Furthermore, depressed patients have reduced levels of FKBP5 after dexamethasone challenge compared with healthy individuals, and that genetic variation in the *FKBP5* gene is associated with the extent of cortisol dysregulation and with the risk of depression and treatment response.[59,90,92] Recently, Binder and colleagues[93] provided evidence for DNA methylation that mediated the combined effects of early-life stress and a genetic polymorphism (rs1360780) on risk of developing psychiatric disorders such as posttraumatic stress disorder (PTSD) and depression. Changes in chromatin structure included regions around distal glucocorticoid response elements and, together with increased cortisol levels (and GR binding), resulted in reductions in DNA methylation in intron 7 of the *FKBP5*gene, which increased *FKBP5* gene transcription in response to GR activation. Overall, these molecular changes resulted in an enhancement of the intracellular ultrashort negative feedback loop regulating GR activity, leading to GR resistance.[93] Interestingly, this appears to be an example where genotype and early trauma can predispose an individual to both PTSD and depression, suggesting that this gene-to-environmental interaction is more generally associated with stress responsiveness and may overlap clinical definitions.

3.2.1.2 Brain-derived neurotrophic factor

Stressors are known to reduce synaptic plasticity in the prefrontal cortex and the hippocampus, two brain regions known that undergo volumetric changes in depressed individuals. In animal models, rats subjected to chronic stress or repeated injections of glucocorticoids show loss of hippocampal neurons and reductions in numbers of dendrites.[48,102] As described above, epigenetic mechanisms may link effects of stress with transcription regulation. One key target gene in mediating synaptic plasticity that responds negatively to stress encodes the neurotrophin, BDNF. Because of its importance in the brain, it is used as an example of a gene-specific target for epigenetic modification.

BDNF is central to regulating the establishment of long-term potentiation (LTP) and synaptic plasticity, the cellular correlates of memory formation. The focus of many laboratories has been to understand activity-dependent mechanisms regulating the BDNF gene expression. Two major activity-dependent promoters have been extensively studied, promoters 1 and 4 (driving exons I and IV, respectively). Promoter 4 has been the focus of many elegant studies showing tight patterns of transcriptional repression and derepression depending on the stimulus provided and the neuronal population.[103–106] Regulation of promoter 1 is also of interest since it drives the expression of a protein isoform that has different functional properties in transfected cell lines and primary neuronal cultures that may also operate *in vivo*.[107–109] Treating rat hippocampal neurons in culture with NMDA treatment induces a durable and robust increase in BDNF gene exon1 mRNAs compared with those containing exon 4.[110] Thus, many studies involving depression models, postmortem analyses, and peripheral cell types as neuronal surrogates have focused on epigenetic mechanisms at these two promoters that apparently show divergent patterns of epigenetic regulation.[105,109]

Regulation of the *BDNF* gene (*BDNF*, human nomenclature; *bdnf*, rodent nomenclature) is complex. In addition to activity-dependent transcriptional regulation, release of BDNF protein is tightly regulated. A common genetic polymorphism, Val66Met, influences activity-dependent release of the mature polypeptide and has a role in cognitive performance.[21] In rats subjected to prolonged (1 month) stress, *bdnf* exon IV mRNA levels were reduced, which correlated with increased levels of CpG methylation in both the dentate gyrus and area CA1 of the hippocampus.[111] In postmortem brains from patients with MDD, levels of BDNF protein and its receptor, TrkB, are decreased in a region-specific manner.[112]

However, BDNF protein levels are elevated in the hippocampi of MDD patients treated with ADs.[113] Related to understanding mechanisms for treating depression, ketamine, an NMDA receptor antagonist with short-term, rapid-acting AD activity, may also target BDNF synthesis. Kavalali and Monteggia[114] showed that the rapid AD action of ketamine is mediated by reduced elongation factor-2 (EF-2) phosphorylation, which acts to release BDNF protein synthesis from suppression. Thus, epigenetic mechanisms have the potential to modify genetic and protein synthesis machinery for the *BDNF* gene and other AD target genes (Fig. 5.3). The use of animal models of depression is essential for gaining mechanistic insight.

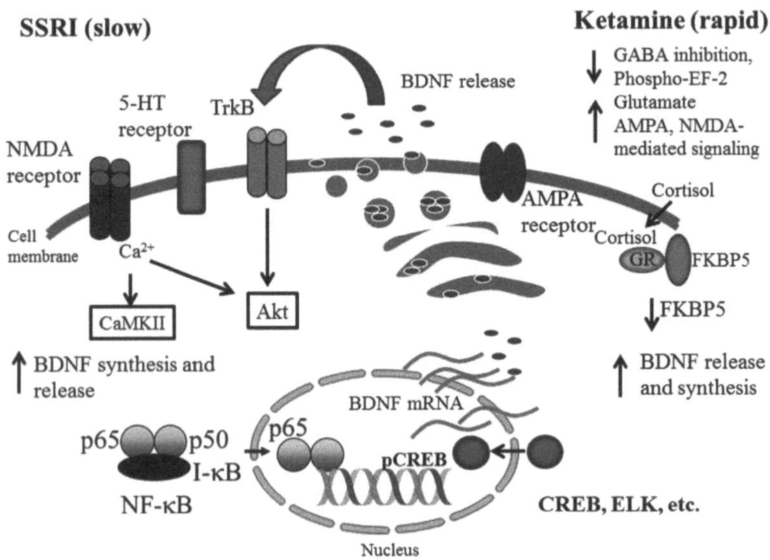

Figure 5.3 An example of rapid- and slow-acting transcriptional and posttranscriptional events by neurons in response to antidepressant (AD) therapy. Epigenetic changes established by AD therapy are expected to be perpetuated in the brain, setting in motion long-term changes in gene expression. This example shows signaling pathways leading to activation-dependent transcription of *BDNF*. Rapid-acting ADs, such as ketamine, by reducing GABA inhibition and promoting compensatory glutamate-mediated signaling, are predicted to increase BDNF protein release within hours followed by transcriptional effects. SSRIs, on the other hand, may require chronic administration because of long-term epigenetic effects that enhance *BDNF* gene expression to stimulate neurogenesis. One of the transcriptional regulatory targets of acute and chronic AD therapy would be cAMP response element-binding (CREB) protein, although other transcription factors also have roles. The effect of chronic stress on *BDNF* expression may be indirect with its normalization by AD treatment acting through DNA demethylation (Ref. 115). (See the color plate.)

In animal models of depression, using electroconvulsive shock (a known treatment for depression) increased levels of histones 3 and 4 (H3 and H4) acetylation at *bdnf* promoters 1 and 4.[116] Increased H3 acetylation and H4 acetylation are regarded as marks of open chromatin and are associated with transcriptional activation. In other studies using social defeat as a stressor, total *bdnf* mRNA was decreased in the hippocampus. Chronic treatment with imipramine normalized *bdnf* mRNA levels following stress[117] and resulted in increases in a histone mark associated with transcriptional activation (H3 acetylation and H3K4me2).[117] The interaction of stress response and changes in epigenetic marks at *bdnf* promoters is underscored by subsequent experiments by other investigators using healthy unstressed mice, where no alteration in H3 acetylation was seen following AD treatment.[118]

While histone modification provides one level of epigenetic regulation, DNA methyltransferases catalyze the addition of methyl groups from *S*-adenosylmethionine (SAMe) to cytosine residues within the context of CpG dinucleotides. DNA methylation has been extensively studied in the central nervous system. DNA methylation patterns are stable and retained in samples of purified DNA, making this epigenetic modification amenable to quantification in archived tissue and blood samples from animal and clinical studies. In addition, the role of DNA methylation in regulating transcription, where increased DNA methylation is generally considered a transcriptionally repressive modification, provides the potential for new approaches to disease treatment.

A central component in DNA methylation is the vitamin folate. Folate serves as a coenzyme in single-carbon transfers in the synthesis of nucleic acids and in the metabolism of amino acids. One of these folate-dependent reactions is in the conversion of homocysteine to methionine in the synthesis of SAMe. Folate deficiency is associated with increased plasma homocysteine and decreased DNA methylation in leukocytes.[119] Thus, a change in the level of folate or the intermediates in SAM synthesis has the potential to influence DNA methylation. The pioneering work of Meaney, Szyf, and coworkers showed that DNA methylation patterns in the brain could be altered in adult mice with methionine.[99] Their experimental paradigm used a maternally induced programming of behavior in neonatal mice that persisted when the mice became adults. When the brains of these adult mice were infused with methionine, their behavior was reprogrammed and this correlated with changes in their HPA responses and patterns of DNA methylation.[99] Taken together, their findings suggested that dietary methionine could be used to influence the epigenetic program in the adult brain. Their experiments provided the conceptual framework for subsequent human studies using folate, methionine, or SAM to alter clinical outcomes.

DNA methylation has a role in global programming of the genome and gene-specific effects. DNA methylation is critical in development and is the molecular process that causes genomic imprinting and X chromosome inactivation. DNA methylation is a substrate for binding of methyl-CpG-binding domain (MBD) proteins. MBD proteins can recruit HDACs and other proteins to condense chromatin and to silence gene expression. One example of gene-specific DNA methylation known to control gene expression involves MeCP2-mediated transcriptional repression of the *bdnf* gene. MeCP2 is a member of the MBD protein family. Its *bdnf* gene-specific binding activity was based on DNA methylation patterns determined in cortical neurons where its action was consistent with a transcriptional repressor that recruits HDACs to condense chromatin.[103,105] Interestingly, inhibition of global DNA methylation by zebularine can restore BDNF expression in an early-life stress model in rats.[111] However, methylation levels at promoter regions of genes and binding of MBD proteins have other functions: MeCP2 also targets DNA by methylation-independent mechanisms that activate transcription,[120,121] which may explain why BDNF is actually reduced at the protein and mRNA levels in mice that lack the *Mecp2* gene.[122] This divergent functionality of MeCP2 emphasizes the importance of substrate specificity in choosing potential therapeutic targets to regulate *BDNF* and possibly other genes in designing potential therapies for depression. This is a topic that will be discussed further in how to consider manipulating the epigenetic program in treating depression.

3.3. Depression-related genes: Epigenome scans

As noted earlier, the estimated heritability of MDD based on twin studies is modest, suggesting that other factors, epigenetic factors, could fill the void. While specific gene surveys as described here for *SLC6A4*, *NR3C1*, *FKBP5*, and *BDNF* have utility based on neurobiological hypotheses, a genome-wide scan to profile epigenetic changes (an epigenome scan) offers the opportunities to discover candidate genes using an unbiased approach and to validate the findings.

The first genome-wide determination of DNA methylation profiles in a community-based sample of adult subjects with MDD was performed using whole blood from 33 individuals reporting a lifetime history of depression and 67 nondepressed individuals.[123] Bisulfite-converted DNA prepared from whole blood was assessed for methylation status at CpG sites in more than 14,000 genes using an Illumina HumanMethylation27 (HM27) BeadChip. Overall, depression-affected individuals had fewer uniquely unmethylated and methylated genes than unaffected control individuals. Computational functional analysis revealed changes in DNA methylation patterns for a number of genes that reflected altered processes in brain development and tryptophan metabolism, with enrichment of the gene having increased methylation in the group with a lifetime history of depression. In contrast, reduced methylation was seen in inflammatory genes IL-6 and CRP, which coincided with increased levels of these proteins in the blood. Interestingly, the findings suggest a role for inflammation in MDD and other comorbid conditions.[124,125] In addition, other genes thought to be associated with the etiology of depression were differentially methylated: *MAOA*, *SLC6A3*, and *GABAR*. Of note is the fact that the Illumina BeadChip did not include CpG sites in the promoter regions of *SLC6A4* or *NR3C1*, so the methylation status of these genes that have been the subject of candidate gene studies could not be evaluated. Some limitations of the study included the fact that this was a cross-sectional study, so changes due to the depression, as opposed to other preexisting conditions, could not be discerned. Also, since whole-blood samples were used, there is the possibility that cell-type heterogeneity could have produced the findings.

The first genome-wide DNA methylation scan on postmortem brain samples from individuals with MDD was performed on 39 frontal cortex samples from MDD subjects compared with 26 healthy control samples.[126] The assay used a methylation-sensitive restriction endonuclease-based approach (methylation-sensitive enzyme *Mcr*BC), implemented on a custom NimbleGen 2.1 M feature mouse CHARM microarray.[126] Of 3.5 million CpGs represented on the array, CHARM identified 224 candidate regions having differences of >10%. Ten of 17 regions were selected for validation. The greatest difference was *PRIMA1* (12–15% increase in MDD upon validation). PRIMA1 anchors acetylcholinesterase in the membrane and may have a role in MDD based on reduced levels in the brain. Consistent with increased DNA methylation, decrease in acetylcholinesterase immunoreactivity was seen in the MDD brain as compared with control, although the result did not reach statistical significance. As a consequence, the findings must be considered suggestive. The scan used only one brain region. It is possible that significant differences in *PRIMA1* DNA methylation levels occur in other brain regions of individuals with MDD. Future studies have focused on increasing the signal-to-noise level of CHARM.

In a recent application of whole-genome RNA sequence analysis (transcriptome profiling), Kohen *et al.*[127] used laser capture microdissection (LCM) to isolate dentate gyrus granule cells in the postmortem human hippocampus in subjects with different neuropsychiatric disorders, including major depression and nonpsychiatric controls. For comparisons relevant to this discussion, transcriptome differences were performed between samples from 17 major depression cases and 29 nonpsychiatric controls. RNA isolated from harvested cells was linearly amplified over two rounds of "aRNA" amplification. Sequencing libraries were prepared and analyzed using Applied Biosystems SOLiD 4, a high-throughput sequencer. Different data normalization methods were used to remove bias against short transcripts. Evidence for reduced signaling of the microRNA *miR*-182 in individuals with major depressions was obtained. Results were validated by showing that the impact of the microRNA on expression of target genes was lost in dentate gyrus granule cells from the brains of subjects who were diagnosed with major depression. *miR*-182 is involved in a number of biological processes, including regulation of circadian rhythm. Disruption of sleep and circadian cycles is a feature of major depression and is consistent with at least one function of this microRNA. While microRNA profiling is relatively new, the identification of specific classes of microRNAs associated with major depression may offer a fresh avenue for designing new therapies for major depression.

3.4. Epigenetic manipulation and depression (global vs. gene-specific effects)

3.4.1 Caveats of epigenetic studies

Epigenetic mechanisms are not isolated events. Instead, these processes interact and influence each other. For example, unmethylated CpG sites at gene promoter regions that are influenced by MBD activity are dependent on the chromatin microenvironment state, which includes histone modifications. Another example is the complex interplay between DNA methylation and histone methylation observed with unmethylated histone H3 (H3K4), which becomes the docking site for DNMTs, resulting in *de novo* DNA methylation and silencing gene transcription.[128]

In addition, epigenetic mechanisms show a high degree of tissue specificity. This presents technical difficulties in obtaining pure populations of cells. With four major cell types interacting in the CNS (neurons, astrocytes, oligodendrocytes, and microglia), this is not a trivial exercise. Overlying cellular specificity are brain region-specific effects. With these issues in mind, in animal models, DNMT inhibitors can impair induction of LTP in hippocampal neurons that are associated with learning deficits.[72] By contrast, DNMT inhibition affects recall but not memory formation. Taken together, these data suggest that circuit-specific patterns of DNA methylation are important in memory formation and maintenance and emphasize the importance of selecting appropriate cell types in determining underlying epigenetic mechanisms.

Related to the selection of appropriate cell types for investigating epigenetic mechanisms, the use of primary cell cultures and methods to enrich populations of cells through cell sorting, or LCM, as described in the previous section, can achieve higher levels of cellular purity. These approaches have clear advantages in reducing confounding signals from other cell types. In addition, the issue of brain cell heterogeneity can be addressed through cell epigenotype-specific marks, which are genomic features that underlie cell type-specific epigenetic variation. The approach can be used to quantify neuronal and glial cells in different brain regions and thereby control for cellular heterogeneity.[129] It can be used to validate and extend studies using DNA methylation profiles to distinguish brain regions.[130]

3.5. Epigenetics of treatment response

Reduced NET has been shown in patients with MDD. However, genetic association findings were inconclusive. The idea of epigenetic control through increased methylation of *SLC6A2* as a mechanism to explain why patients with MDD have a dysfunctional NET is appealing. A recent study was undertaken to characterize the DNA methylation state of the *SLC6A2* gene promoter in patients with untreated MDD and panic disorder before and after treatment.[131] DNA was obtained from blood leukocytes of 36 untreated MDD patients, 36 panic disorder subjects, and 70 healthy controls. DNA was also obtained from five MDD and four panic disorder subjects before and after treatment. Treatment of MDD patients was for 3 months. Two regions of the *SLC6A2* promoter were analyzed between −515 and +167 relative to the transcription start site. Promoter methylation was investigated by bisulfite sequencing and EpiTYPER methylation analysis (MassARRAY, Sequenom). MDD patients did not differ significantly from healthy controls in the pattern of methylation for two regions of the *SLC6A2* promoter analyzed. In addition, no significant changes were seen in response to SSRI treatment. In general, the regions analyzed in the *SLC6A2* promoter were hypomethylated. Whether other epigenetic marks in this promoter region are altered in MDD or by SSRI treatment remains to be determined.

3.6. Can epigenetic manipulation improve clinical outcome in major depression?

Epigenetic modulations can potentially affect AD treatment by modulating the expression of genes involved in the pharmacokinetics and pharmacodynamics of ADs.[132] Identification of genetic biomarkers could facilitate the choice of treatment and prediction of the AD response, thus having a potential to improve clinical outcome in MDD by way of "personalized medicine." However, current literature is inconclusive, mostly with small effect sizes and lack of replication that is required before applying to real-world clinical scenarios.

Yu *et al.*[133] demonstrated that desipramine, but not fluoxetine, has AD effects on *bdnf* (+/Met66) mice, suggesting that specific classes of AD may be a more effective treatment option for depressive symptoms in humans with the Met66 variant of the *BDNF* gene. The data from GWAS and candidate gene studies exemplify the complexity of the field and the need for more studies, as there is unlikely a single model of epigenetic modulation that could lead to improved outcomes.

Folate is associated with DNA methylation and has a potential to modulate epigenetic programming. Vitamin B12, or cobalamin, is involved in DNA synthesis and regulation, in addition to fatty acid synthesis and energy metabolism. Omega fatty acids have an essential role in neurodevelopment. Docosahexaenoic acid (DHA) also influences neurotrophins, including BDNF. Dhobale and Joshi hypothesize that during pregnancy, altered maternal micronutrients, such as folic acid, vitamin B12, and DHA, may cause epigenetic modifications of neurotrophins, contributing to the risk for neurodevelopmental disorders in children born preterm.[134] Toffoli *et al.*[135] conducted a study in a rat's offspring whose mother was exposed to fluoxetine and found that the combined treatment with both fluoxetine and folate induced an increase in the methylation of the hippocampus indicating the potential of folic acid to modulate this epigenetic alteration.

The use of folate preparations especially the 5-MTHF (L-methylfolate) formulation has shown efficacy as adjunctive therapy or monotherapy in reducing depressive symptoms in patients with normal and low folate levels.[136,137] Later, Papakostas *et al.*[138] showed that in a subset of SSRI-resistant MDD patients who have specific biomarkers associated with inflammation or metabolism and genomic markers associated with

L-methylfolate synthesis and metabolism, treatment with L-methylfolate was superior compared to placebo; however, confirmatory studies are needed.

As noted earlier, DNA methylation patterns in the brain could be altered with methionine, thus providing the conceptual framework for subsequent human studies using folate, methionine, or SAM to alter clinical outcomes by influencing epigenetic programming in the brain. SAMe, a naturally occurring substance, has been available as an alternative treatment for depression for nearly 30 years. Studies support efficacy of SAMe as an augmentation agent in treatment-resistant MDD.[139,140] A recent RCT study of a subsample (from a single site) of a previous two-site study investigating AD efficacy of SAMe in MDD versus escitalopram and placebo provided encouraging evidence for the use of SAMe as a monotherapy in the treatment of MDD.[141] However, given the paucity of evidence and prior negative studies of oral SAMe in depression,[142] more controlled studies with perhaps larger "real-world" clinical samples and longer follow-up periods are needed.

Compared with other forms of folates, 5-methyltetrahydrofolate (L-methylfolate or 5-MTHF) may represent a preferable treatment option for MDD given its greater bioavailability in patients with a genetic polymorphism and the lower risk of specific side effects associated with folic acid.[143]

A recent literature review by Gvozdic *et al.*[144] on pharmacogenetics of AD response concluded that, despite promising findings for several variants in candidate genes involved, contradictory results would limit clinical utility. Future studies should focus on comprehensive functional biomarker analyses with consideration of gene-to-gene and gene-to-environment interactions. The answers to these questions could be the key to how we treat depression in the future. How do we develop these interventions and for whom? Based on the topics covered in this chapter, the answer is likely not a "one size fits all" approach. Knowledge of one's epigenotype will go a long way to understanding disease processes and guide development of new therapies.

Much of what we know of drugs that act on the epigenome comes from the cancer field, where they are generally classified as DNA methyltransferase inhibitors and HDAC inhibitors. These classes of drugs have already made their way into clinical settings, treating several cancer types (Clinical Trials.gov). However, because of the broad and nonspecific nature of epigenetic-modifying drugs and their adverse effects, more targeted epigenetic modifiers are currently under intense research investigation and are just entering the clinical trials pipeline.

As noted earlier in this chapter, the etiologically of complex neuropsychiatric disorders is far from complete, so research into a new generation of epigenetic-modifying drugs for treating MDD is still at an early stage of development. A challenge will be to understand how epigenetic marks are induced, maintained, or removed: their manipulation at specific cellular target sites in neurons and glia is still unrealized. For epigenetic drugs to act with high specificity and without having many adverse effects, these mechanisms must be well understood.

ACKNOWLEDGMENTS AND DISCLOSURES

The authors wish to recognize the participants of the STAR*D study, without whom much of this work could not have been possible. The authors declare no conflict of interest.

REFERENCES

1. American Psychiatric Association. *Diagnostic and Statistical Manual of Mental Disorders*. 5th ed. Arlington, VA: American Psychiatric Publishing; 2013.
2. Uher R, Payne JL, Pavlova B, Perlis RH. Major depressive disorder in DSM-5: implications for clinical practice and research of changes from DSM-IV. *Depress Anxiety*. 2013;31:459–471.
3. Uher R. Genomics and the classification of mental illness: focus on broader categories. *Genome Med*. 2013;5:97.
4. Whiteford HA, Degenhardt L, Rehm J, et al. Global burden of disease attributable to mental and substance use disorders: findings from the Global Burden of Disease Study 2010. *Lancet*. 2013;382:1575–1586.
5. Murray CJL, Lopez AD. *The Global Burden of Disease: A Comprehensive Assessment of Mortality and Disability from Diseases, Injuries and Risk Factors in 1990 and Projected to 2020*. Geneva, Switzerland: World Health Organization; 1996.
6. Kessler RC, Bromet EJ. The epidemiology of depression across cultures. *Annu Rev Public Health*. 2013;34:119–138.
7. Kessler RC, Berglund PA, Demler O, Jin R, Walters EE. Lifetime prevalence and age-of-onset distributions of DSM-IV disorders in the National Comorbidity Survey Replication (NCS-R). *Arch Gen Psychiatry*. 2005;62:593–602.
8. Bromet E, Andrade LH, Hwang I, et al. Cross-national epidemiology of DSM-IV major depressive episode. *BMC Med*. 2011;9:90.
9. Dalva MB, McClelland AC, Kayser MS. Cell adhesion molecules: signaling functions at the synapse. *Nat Rev Neurosci*. 2007;8:206–220.
10. El-Hage W, Leman S, Camus V, Belzung C. Mechanisms of antidepressant resistance. *Front Pharmacol*. 2013;4:146.
11. Shores MM, Kivlahan DR, Sadak TI, Li EJ, Matsumoto AM. A randomized, double-blind, placebo-controlled study of testosterone treatment in hypogonadal older men with subthreshold depression (dysthymia or minor depression). *J Clin Psychiatry*. 2009;70:1009–1016.
12. Cohen LS, Soares CN, Vitonis AF, Otto MW, Harlow BL. Risk for new onset of depression during the menopausal transition: the Harvard study of moods and cycles. *Arch Gen Psychiatry*. 2006;63:385–390.
13. Goel N, Bale TL. Examining the intersection of sex and stress in modeling neuropsychiatric disorders. *J Neuroendocrinol*. 2009;21:415–420.

14. Kessler RC. The costs of depression. *Psychiatr Clin North Am*. 2012;35:1–14.
15. Pozuelo L, Tesar G, Zhang J, Penn M, Franco K, Jiang W. Depression and heart disease: what do we know, and where are we headed? *Cleve Clin J Med*. 2009;76:59–70.
16. Sher Y, Lolak S, Maldonado JR. The impact of depression in heart disease. *Curr Psychiatry Rep*. 2010;12:255–264.
17. Charney DS. Monoamine dysfunction and the pathophysiology and treatment of depression. *J Clin Psychiatry*. 1998;14:11–14.
18. Krishnan V, Nestler EJ. Linking molecules to mood: new insight into the biology of depression. *Am J Psychiatry*. 2010;167:1305–1320.
19. Trivedi MH, Rush AJ, Wisniekski SR, et al. Evaluation of outcomes with citalopram for depression using measurement-based care in STAR*D: implications for clinical practice. *Am J Psychiatry*. 2006;163:28–40.
20. Murrough JW, Iosifescu DV, Chang LC, et al. Antidepressant efficacy of ketamine in treatment-resistant major depression: a two-site randomized controlled trial. *Am J Psychiatry*. 2013;170:1134–1142.
21. Egan MF, Kojima M, Callicott JH, et al. The BDNF val66met polymorphism affects activity-dependent secretion of BDNF and human memory and hippocampal function. *Cell*. 2003;112:257–269.
22. Laje G, Lally N, Mathews D, et al. Brain-derived neurotrophic factor Val66Met polymorphism and antidepressant efficacy of ketamine in depressed patients. *Biol Psychiatry*. 2012;72:e27–e28.
23. Lipsky RH, Marini AM. Brain-derived neurotrophic factor in neuronal survival and behavior-related plasticity. *Ann N Y Acad Sci*. 2007;1122:130–143.
24. Neumeister A, Yuan P, Yound TA, et al. Effects of tryptophan depletion on serum levels of brain-derived neurotrophic factor in unmedicated patients with remitted depression and healthy subjects. *Am J Psychiatry*. 2005;162:805–807.
25. Mann JJ, Currier D. Effects of genes and stress on the neurobiology of depression. *Int Rev Neurobiol*. 2006;73:153–189.
26. Hayley S, Litteljohn D. Neuroplasticity and the next wave of antidepressant strategies. *Front Cell Neurosci*. 2013;7:218.
27. Shimizu E, Hashimoto K, Okamura N, et al. Alterations of serum levels of brain-derived neurotrophic factor (BDNF) in depressed patients with or without antidepressants. *Biol Psychiatry*. 2003;54:70–75.
28. Sun H, Kennedy PJ, Nestler EJ. Epigenetics of the depressed brain: role of histone acetylation and methylation. *Neuropsychopharmacology*. 2013;38:124–137.
29. Dedoni S, Olianas MC, Ingianni A, Onali P. Type I interferons impair BDNF-induced cell signaling and neurotrophic activity in differentiated human SH-SY5Y neuroblastoma cells and mouse primary cortical neurons. *J Neurochem*. 2012;122:58–71.
30. Berton O, McClung CA, Dileone RJ, et al. Essential role of BDNF in the mesolimbic dopamine pathway in social defeat stress. *Science*. 2006;31:864–868.
31. Martocchia A, Curto M, Scaccianoce S, et al. Effects of escitalopram on serum BDNF levels in elderly patients with depression: a preliminary report. *Aging Clin Exp Res*. 2014;26(4):461–464.
32. Pittenger C, Duman RS. Stress, depression, and neuroplasticity: a convergence of mechanisms. *Neuropsychopharmacology*. 2008;33:88–109.
33. Krishnan V, Nestler EJ. The molecular neurobiology of depression. *Nature*. 2008;455:894–902.
34. Wainwright SR, Galea LA. The neural plasticity theory of depression: assessing the roles of adult neurogenesis and PSA-NCAM within the hippocampus. *Neural Plast*. 2013;2013:805497.
35. Holsboer F. The corticosteroid receptor hypothesis of depression. *Neuropsychopharmacology* 2000;23:477–501.

36. Henckens MJ, van Wingen GA, Joëls M, Fernández G. Time-dependent corticosteroid modulation of prefrontal working memory processing. *Proc Natl Acad Sci USA*. 2011;108:5801–5806.
37. Yuen EY, Liu W, Karatsoreos IN, Feng J, McEwen BS, Yann Z. Acute stress enhances glutamatergic transmission in prefrontal cortex and facilitates working memory. *Proc Natl Acad Sci USA*. 2009;106:14075–14079.
38. Crowley GT, Khatami M, Vasavada N. The dexamethasone suppression test in patients with mood disorders. *J Clin Psychiatry*. 1996;57:470–484.
39. Mellsop GW, Hutton JD, Delahunt JW. Dexamethasone suppression test as a simple measure of stress? *Br Med J (Clin Res Ed)*. 1985;290:1804–1806.
40. Brown WA, Shuey I. Response to dexamethasone and subtype of depression. *Arch Gen Psychiatry*. 1980;37:747–751.
41. Rush AJ, Giles DE, Schlesser MA, et al. The dexamethasone suppression test in patients with mood disorders. *J Clin Psychiatry*. 1996;57:470–484.
42. Behnken A, Bellingrath S, Symanczik JP, et al. Associations between cognitive performance and cortisol reaction to the DEX/CRH test in patients recovered from depression. *Psychoneuroendocrinology*. 2013;38:447–454.
43. Wilkinson CW, Peskind ER, Raskind MA. Decreased hypothalamic-pituitary-adrenal axis sensitivity to cortisol feedback inhibition in human aging. *Neuroendocrinology*. 1997;65:79–90.
44. Zovato S, Simoncini M, Gottardo C, et al. Dexamethasone suppression test: corticosteroid receptors regulation in mononuclear leukocytes in young and aged subjects. *Aging*. 1996;8:360–364. http://www.who.int/mediacentre/factsheets/fs369/en/ [accessed December, 2013].
45. Colla M, Kronenberg G, Deuschle M, et al. Hippocampal volume reduction and HPA-system activity in major depression. *J Psychiatr Res*. 2007;41:553–560.
46. MacQueen GM, Yucel K, Taylor VH, Macdonald K, Joffe R. Posterior hippocampal volumes are associated with remission rates in patients with major depressive disorder. *Biol Psychiatry*. 2008;64:880–883.
47. Drevets WC, Price JL, Furey ML. Brain structural and functional abnormalities in mood disorders: implications for neurocircuitry models of depression. *Brain Struct Funct*. 2008;213:93–118.
48. Sapolsky RM. Glucocorticoids and hippocampal atrophy in neuropsychiatric disorders. *Arch Gen Psychiatry*. 2000;57:925–935.
49. Frodl T, O'Keane V. How does the brain deal with cumulative stress? A review with focus on developmental stress, HPA axis function and hippocampal structure in humans. *Neurobiol Dis*. 2013;52:24–37.
50. Kelly WF, Kelly MJ, Faragher B. A prospective study of psychiatric and psychological aspects of Cushing's syndrome. *Clin Endocrinol (Oxf)*. 1996;45:715–720.
51. Ragnarsson O, Berglund P, Eder DN, Hohannsson G. Long-term cognitive impairments and attention deficits in patients with Cushing's disease and cortisol-producing adrenal adenoma in remission. *J Clin Endocrinol Metab*. 2012;97:E1640–E1648.
52. Ising M, Horstmann S, Kloiber S, et al. Combined dexamethasone/corticotropin releasing hormone test predicts treatment response in major depression—a potential biomarker? *Biol Psychiatry*. 2007;62:47–54.
53. Snyder JS, Soumier A, Brewer M, Pickel J, Cameron HA. Adult hippocampal neurogenesis buffers stress responses and depressive behaviour. *Nature*. 2011;476:458–461.
54. Van Kampen M, Kramer M, Hiemke C, Flügge G, Fuchs E. The chronic psychosocial stress paradigm in male tree shrews: evaluation of a novel animal model for depressive disorders. *Stress*. 2002;5:37–46.
55. Fuchs E, Flügge G. Social stress in tree shrews: effects on physiology, brain function, and behavior of subordinate individuals. *Pharmacol Biochem Behav*. 2002;7:247–258.

56. American Psychiatric Association. *APA Practice Guideline for the Treatment of Patients with Major Depressive Disorder*. 3rd ed. Arlington, Virginia: American Psychiatric Association; 2013. http://www.psychiatryonline.com/pracGuide/pracGuideTopic_7.aspx [accessed December, 2013].
57. Rush AJ, Trivedi MH, Wisniewski SR, et al. Acute and longer-term outcomes in depressed outpatients requiring one or several treatment steps: a STAR*D report. *Am J Psychiatry*. 2006;163:1905–1917.
58. Lam RW. Onset, time course and trajectories of improvement with antidepressants. *Eur Neuropsychopharmacol*. 2012;22(Suppl. 3):S492–S498.
59. Menke A, Klengel T, Binder EB. Epigenetics, depression and antidepressant treatment. *Curr Pharm Des*. 2012;18:5879–5889.
60. Tsankova N, Renthal W, Kumar A, Nestler EJ. Epigenetic regulation in psychiatric disorders. *Nat Rev Neurosci*. 2007;8:355–367.
61. El Hage W, Powell JF, Surguladze SA. Vulnerability to depression: what is the role of stress genes in gene x environment interaction? *Psychol Med*. 2009;39: 1407–1411.
62. Santarelli L, Saxe M, Gross C, et al. Requirement of hippocampal neurogenesis for the behavioral effects of antidepressants. *Science*. 2003;301:805–809.
63. Bessa JM, Ferreira D, Melo I, et al. The mood-improving actions of antidepressants do not depend on neurogenesis but are associated with neuronal remodeling. *Mol Psychiatry*. 2009;14:764–773.
64. Tang SW, Helmeste D, Leonard B. Is neurogenesis relevant in depression and in the mechanism of antidepressant drug action? A critical review. *World J Biol Psychiatry*. 2012;13:402–412.
65. Perera TD, Coplan JD, Lisanby SH, et al. Antidepressant-induced neurogenesis in the hippocampus of adult nonhuman primates. *J Neurosci*. 2007;27:4894–4901.
66. McGuffin P, Cohen S, Knight J. Homing in on depression genes. *Am J Psychiatry*. 2007;164:195–197.
67. Shyn SI, Hamilton SP. The genetics of major depression: moving beyond the monoamine hypothesis. *Psychiatr Clin North Am*. 2010;33:125–140.
68. Shyn SI, Shi J, Kraft JB, et al. Novel loci for major depression identified by genome-wide association study of Sequenced Treatment Alternatives to Relieve Depression and meta-analysis of three studies. *Mol Psychiatry*. 2011;16:202–215.
69. Crisafulli C, Fabbri C, Porcelli S, et al. Pharmacogenetics of antidepressants. *Front Pharmacol*. 2011;2:6.
70. Hu X, Lipsky RH, Zhu G, et al. Gain-of-function genotypes are linked to obsessive-compulsive disorder. *Am J Hum Genet*. 2006;78:815–826.
71. Lesch KP, Bengel D, Heils A, et al. Association of anxiety-related traits with a polymorphism in the serotonin transporter gene regulatory region. *Science*. 1996;2(74): 1527–1531.
72. Levinson DF. The genetics of depression: a review. *Biol Psychiatry*. 2006;60:84–92.
73. Hu X-Z, Rush AJ, Charney D, et al. Association between a functional serotonin transporter promoter polymorphism and citalopram treatment in adult outpatients with major depression. *Arch Gen Psychiatry*. 2007;64:783–792.
74. McMahon FJ, Buervenich S, Charney D, et al. Variation in the gene encoding the serotonin 2A receptor is associated with outcome of antidepressant treatment. *Am J Hum Genet*. 2006;78:804–814.
75. Mrazek DA, Rush AJ, Biernacka JM, et al. SLC6A4 variation and citalopram response. *Am J Med Genet B Neuropsychiatr Genet*. 2009;150B:341–351.
76. Karg K, Burmeister M, Shedden K, Sen S. The serotonin transporter promoter variant (5-HTTLPR), stress, and depression meta-analysis revisited: evidence of genetic moderation. *Arch Gen Psychiatry*. 2011;68:444–454.

77. Hu X, Tian F, Ecklund JM, Lipsky RH. Differential responsiveness of the serotonin transporter gene in pharmacologic treatment of adult depression: potential for understanding epigenetic-genetic interactions. In: *Gordon Research Conference, Human Genetics and Genomics July 17–22*; 2011.
78. Dudbridge F, Gusnanto A. Estimation of significance thresholds for genomewide association scans. *Genet Epidemiol.* 2008;32:227–234.
79. Major Depressive Disorder Working Group of the Psychiatric GWAS Consortium, Ripke S, Wray NR, et al. A mega-analysis of genome-wide association studies for major depressive disorder. *Mol Psychiatry.* 2013;18:497–511.
80. Singh AB, Bousman CA, Ng C, Berk M. Antidepressant pharmacogenetics. *Curr Opin Psychiatry.* 2014;27:43–51.
81. Shi Y, Yuan Y, Xu Z, et al. Genetic variation in the calcium/calmodulin-dependent protein kinase (CaMK) pathway is associated with antidepressant response in females. *J Affect Disord.* 2012;136:558–566.
82. Singh AB, Bousman CA, Ng CH, Byron K, Berk M. ABCB1 polymorphism predicts escitalopram dose needed for remission in major depression. *Transl Psychiatry.* 2012;2: e198.
83. Laje G, McMahon FJ. Genome-wide association studies of antidepressant outcome: a brief review. *Prog Neuropsychopharmacol Biol Psychiatry.* 2011;35:1553–1557.
84. Garriock HA, Kraft JB, Shyn SI, et al. A genomewide association study of citalopram response in major depressive disorder. *Biol Psychiatry.* 2010;167:133–138.
85. Ising M, Lucae S, Binder EB, et al. A genomewide association study points to multiple loci that predict antidepressant drug treatment outcome in depression. *Arch Gen Psychiatry.* 2009;66:966–975.
86. Uher R, Perroud N, Ng MY, et al. Genome-wide pharmacogenetics of antidepressant response in the GENDEP project. *Am J Psychiatry.* 2010;167:555–564.
87. Serretti A, Kato M, De Ronchi D, Kinoshita T. Meta-analysis of serotonin transporter gene promoter polymorphism (5-HTTLPR) association with selective serotonin reuptake inhibitor efficacy in depressed patients. *Mol Psychiatry.* 2007;12(3):247–257.
88. Taylor MJ, Sen S, Bhagwagar Z. Antidepressant response and the serotonin transporter gene-linked polymorphic region. *Biol Psychiatry.* 2010;68:536–543.
89. Porcelli S, Fabbri C, Serretti A. Meta-analysis of serotonin transporter gene promoter polymorphism (5-HTTLPR) association with antidepressant efficacy. *Eur Neuropsychopharmacol.* 2012;22(4):239–258.
90. Binder EB, Salyakina D, Lichtner P, et al. Polymorphisms in FKBP5 are associated with increased recurrence of depressive episodes and rapid response to antidepressant treatment. *Nat Genet.* 2004;36:1319–1325.
91. Binder EB. The role of FKBP5, a co-chaperone of the glucocorticoid receptor in the pathogenesis and therapy of affective and anxiety disorders. *Psychoneuroendocrinology.* 2009;34(Suppl. 1):S186–S195.
92. Lekman M, Laje G, Charney D, et al. The FKBP5-gene in depression and treatment response—an association study in the Sequenced Treatment Alternatives to Relieve Depression (STAR*D) Cohort. *Biol Psychiatry.* 2008;63:1103–1110.
93. Klengel T, Mehta D, Anacker C, et al. Allele-specific FKBP5 DNA demethylation mediates gene-childhood trauma interactions. *Nat Neurosci.* 2013;16:33–41.
94. Sullivan PF, Neale MC, Kendler KS. Genetic epidemiology of major depression: review and meta-analysis. *Am J Psychiatry.* 2000;157:1552–1562.
95. Bonanno GA, Galea S, Bucciarelli A, Vlahov D. What predicts psychological resilience after disaster? The role of demographics, resources, and life stress. *J Consult Clin Psychol.* 2007;7:671–682.
96. Fossion P, Leys C, Kempenaers C, Braun S, Verbanck P, Linkowski P. Depression, anxiety and loss of resilience after multiple traumas: an illustration of a mediated

moderation model of sensitization in a group of children who survived the Nazi Holocaust. *J Affect Disord.* 2013;151:973–979.
97. Caspi A, Sugden K, Moffitt TE, et al. Influence of life stress on depression: moderation by a polymorphism in the 5-HTT gene. *Science.* 2003;18:386–389.
98. Risch N, Herrell R, Lehner T, et al. Interaction between the serotonin transporter gene (5-HTTLPR), stressful life events, and risk of depression: a meta-analysis. *JAMA.* 2009;301:2462–2471.
99. Weaver IC, Cervoni N, Champagne FA, et al. Epigenetic programming by maternal behavior. *Nat Neurosci.* 2004;7:847–854.
100. Fraga MF, Ballestar E, Paz MF, et al. Epigenetic differences arise during the lifetime of monozygotic twins. *Proc Natl Acad Sci USA.* 2005;102:10604–10609.
101. Ruhe HG, Mason NS, Schene AH. Mood is indirectly related to serotonin, norepinephrine, and dopamine levels in humans: a meta-analysis of monoamine depletion studies. *Mol Psychiatry.* 2007;12:331–359.
102. Watanabe Y, Gould E, McEwen BS. Stress induces atrophy of apical dendrites of hippocampal CA3 pyramidal neurons. *Brain Res.* 1992;588:341–345.
103. Chen WG, Chang Q, Lin Y, et al. Derepression of BDNF transcription involves calcium-dependent phosphorylation of MeCP2. *Science.* 2003;302:885–889.
104. Lipsky RH, Xu K, Zhu D, et al. Nuclear factor kappaB is a critical determinant in N-methyl-D-aspartate receptor-mediated neuroprotection. *J Neurochem.* 2001;78: 254–264.
105. Martinowich K, Hattor I, Wu H, et al. DNA methylation-related chromatin remodeling in activity-dependent BDNF gene regulation. *Science.* 2003;302:890–893.
106. Timmusk T, Belluardo N, Persson H, Metsis M. Developmental regulation of brain-derived neurotrophic factor messenger RNAs transcribed from different promoters in the rat brain. *Neuroscience.* 1994;60:287–291.
107. Jiang X, Zhou J, Mash DC, Marini AM, Lipsky RH. Human BDNF isoforms are differentially expressed in cocaine addicts and are sorted to the regulated secretory pathway independent of the Met66 substitution. *Neuromol Med.* 2009;11:1–12.
108. Liu QR, Lu L, Zhu XG, Gong JP, Shaham Y, Uhl GR. Rodent BDNF genes, novel promoters, novel splice variants, and regulation by cocaine. *Brain Res.* 2006;1067:1–12.
109. Marini AM, Jiang X, Wu X, et al. Role of brain-derived neurotrophic factor and NF-kappaB in neuronal plasticity and survival: from genes to phenotype. *Restor Neurol Neurosci.* 2004;22:121–130.
110. Lipsky RH. Epigenetic mechanisms regulating learning and long-term memory. *Int J Dev Neurosci.* 2013;31:353–358.
111. Roth TL, Lubin FD, Funk AJ, Sweatt JD. Lasting epigenetic influence of early-life adversity on the BDNF gene. *Biol Psychiatry.* 2009;65:760–769.
112. Kozicz T, Tilburg-Ouwens D, Faludi G, Palkovits M, Roubos E. Gender-related urocortin 1 and brain-derived neurotrophic factor expression in the adult human midbrain of suicide victims with depression. *Neuroscience.* 2008;152: 1015–1023.
113. Chen B, Dowlatshahi D, MacQueen GM, Wang JF, Young LT. Increased hippocampal BDNF immunoreactivity in subjects treated with antidepressant medication. *Biol Psychiatry.* 2001;50:260–265.
114. Kavalali ET, Monteggia LM. Synaptic mechanisms underlying rapid antidepressant action of ketamine. *Am J Psychiatry.* 2012;169:1150–1156.
115. Guiddotti G, Calabrese F, Anacker C, Racagni G, Pariante CM, Riva MA. Glucocorticoid receptor and FKBP5 expression is altered following exposure to chronic stress: modulation by antidepressant treatment. *Neuropsychopharmacology.* 2013;38:616–627.

116. Tsankova NM, Kumar A, Nestler EJ. Histone modifications at gene promoter regions in rat hippocampus after acute and chronic electroconvulsive seizures. *J Neurosci.* 2004;24:5603–5610.
117. Tsankova NM, Berton O, Renthal W, Kumar A, Neve RL, Nestler EJ. Sustained hippocampal chromatin regulation in a mouse model of depression and antidepressant action. *Nat Neurosci.* 2006;9:519–525.
118. Karpova NN, Lindolm J, Prunnsild P, Timmusk T, Castren E. Long-lasting behavioural and molecular alterations induced by early postnatal fluoxetine exposure are restored by chronic fluoxetine treatment in adult mice. *Eur Neuropsychopharmacol.* 2009;19:97–108.
119. Farah A. The role of L-methylfolate in depressive disorders. *CNS Spectr.* 2009;14:2–7.
120. Baubec T, Ivánek R, Lienert F, Schübeler D. Methylation-dependent and -independent genomic targeting principles of the MBD protein family. *Cell.* 2013;153:480–492.
121. Chahrour M, Jung SY, Shaw C, et al. MeCP2, a key contributor to neurological disease, activates and represses transcription. *Science.* 2008;320:1224–1229.
122. Chang Q, Khare G, Dani V, Nelson S, Jaenisch R. The disease progression of Mecp2 mutant mice is affected by the level of BDNF expression. *Neuron.* 2006;49:341–348.
123. Uddin M, Koenen KC, Aiello AE, Wildman DE, de los Santos R, Galea S. Epigenetic and inflammatory marker profiles associated with depression in a community-based epidemiologic sample. *Psychol Med.* 2011;41:997–1007.
124. Gold SM, Irwin MR. Depression and immunity: inflammation and depressive symptoms in multiple sclerosis. *Immunol Allergy Clin North Am.* 2009;29:309–320.
125. Miller AH, Maletic V, Raison CL. Inflammation and its discontents: the role of cytokines in the pathophysiology of major depression. *Biol Psychiatry.* 2009;65:732–741.
126. Sabunciyan S, Aryee MJ, Irizarry RA, et al. Genome-wide DNA methylation scan in major depressive disorder. *PLoS One.* 2012;7:e34451.
127. Kohen R, Dobra A, Tracy JH, Haugen E. Transcriptome profiling of human hippocampus dentate gyrus granule cells in mental illness. *Transl Psychiatry.* 2014;4:e366.
128. Szyf M. Epigenetics, DNA methylation, and chromatin modifying drugs. *Annu Rev Pharmacol Toxicol.* 2009;49:243–263.
129. Guintivano J, Aryee MJ, Kaminsky ZA. A cell epigenotype specific model for the correction of brain cellular heterogeneity bias and its application to age, brain region and major depression. *Epigenetics.* 2013;8:290–302.
130. Ladd-Acosta C, Pevsner J, Sabunciyan S, et al. DNA methylation signatures within the human brain. *Am J Hum Genet.* 2007;81:1304–1315.
131. Bayles R, Baker EK, Jowett JBM, Barton D, Esler M, El-Osta A, Lambert G. Methylation of the SLC6a2 gene promoter in major depression and panic disorder. *PLoS One.* 2012;8:e83223.
132. Archer T, Oscar-Berman M, Blum K, Gold M. Epigenetic modulation of mood disorders. *J Genet Syndr Gene Ther.* 2013;11:4(120).
133. Yu H, Wang DD, Wang Y, Liu T, Lee FS, Chen ZY. Variant brain-derived neurotrophic factor Val66Met polymorphism alters vulnerability to stress and response to antidepressants. *J Neurosci.* 2012;32:4092–4101.
134. Dhobale M, Joshi S. Altered maternal micronutrients (folic acid, vitamin B(12)) and omega 3 fatty acids through oxidative stress may reduce neurotrophic factors in preterm pregnancy. *J Matern Fetal Neonatal Med.* 2012;25:317–323.
135. Toffoli LV, Rodrigues Jr GM, Oliveira JF, et al. Maternal exposure to fluoxetine during gestation and lactation affects the DNA methylation programming of rat's offspring: modulation by folic acid supplementation. *Behav Brain Res.* 2014;265: 142–147.

136. Fava M, Mischoulon D. Folate in depression: efficacy, safety, differences in formulations, and clinical issues. *J Clin Psychiatry*. 2009;70:12–17.
137. Morris DW, Trivedi MH, Rush AJ. Folate and unipolar depression. *J Altern Complement Med*. 2008;14:277–285.
138. Papakostas GI, Shelton RC, Zajecka JM, et al. Effect of adjunctive L-methylfolate 15 mg among inadequate responders to SSRIs in depressed patients who were stratified by biomarker levels and genotype: results from a randomized clinical trial. *J Clin Psychiatry*. 2014;75:855–863.
139. De Berardis D, Marini S, Serroni N, et al. S-Adenosyl-L-methionine augmentation in patients with stage II treatment-resistant major depressive disorder: an open label, fixed dose, single-blind study. *ScientificWorldJournal*. 2013;2013:204649.
140. Papakostas GI, Mischoulon D, Shyu I, Alpert JE, Fava M. S-Adenosyl methionine (SAMe) augmentation of serotonin reuptake inhibitors for antidepressant nonresponders with major depressive disorder: a double-blind, randomized clinical trial. *Am J Psychiatry*. 2010;167:942–948.
141. Sarris J, Papakostas GI, Vitolo O, Fava M, Mischoulon D. S-Adenosyl methionine (SAMe) versus escitalopram and placebo in major depression RCT: efficacy and effects of histamine and carnitine as moderators of response. *J Affect Disord*. 2014;164:76–81.
142. Papakostas GI. Evidence for S-adenosyl-L-methionine (SAM-e) for the treatment of major depressive disorder. *J Clin Psychiatry*. 2009;70(Suppl. 5):18–22.
143. Papakostas GI, Cassiello CF, Iovieno N. Folates and S-adenosylmethionine for major depressive disorder. *Can J Psychiatry*. 2012;57:406–413.
144. Gvozdic K, Brandl EJ, Taylor DL, Müller DJ. Genetics and personalized medicine in antidepressant treatment. *Curr Pharm Des*. 2012;18:5853–5878.

CHAPTER SIX

Epigenetic Factors in Intellectual Disability: The Rubinstein–Taybi Syndrome as a Paradigm of Neurodevelopmental Disorder with Epigenetic Origin

Jose P. Lopez-Atalaya, Luis M. Valor, Angel Barco
Instituto de Neurociencias, Universidad Miguel Hernández-Consejo Superior de Investigaciones Científicas, Alicante, Spain

Contents

Progress in Molecular Biology and Translational Science, Volume 128
ISSN 1877-1173
http://dx.doi.org/10.1016/B978-0-12-800977-2.00006-1

Abstract

The number of genetic syndromes associated with intellectual disability that are caused by mutations in genes encoding chromatin-modifying enzymes has sharply risen in the last decade. We discuss here a neurodevelopmental disorder, the Rubinstein–Taybi syndrome (RSTS), originated by mutations in the genes encoding the lysine acetyltransferases CBP and p300. We first describe clinical and genetic aspects of the syndrome to later focus on the insight provided by the research in animal models of this disease. These studies have not only clarified the molecular etiology of RSTS and helped to dissect the developmental and adult components of the syndrome but also contributed to outline some important connections between epigenetics and cognition. We finally discuss how this body of research has opened new venues for the therapeutic intervention of this currently untreatable disease and present some of the outstanding questions in the field. We believe that the progress in the understanding of this rare disorder also has important implications for other intellectual disability disorders that share an epigenetic origin.

1. INTELLECTUAL DISABILITY AND CHROMATIN-MODIFYING ENZYMES

Numerous genes have been associated with neurodevelopmental disorders in which intellectual capabilities are diminished (from now on referred to as intellectual disability disorders or IDDs). This group includes more than 50 genes (~10% of the total number of genes directly linked to IDDs) that encode proteins involved in the epigenetic editing of the genome and other components of chromatin-remodeling complexes.[1] This represents an enrichment for chromatin-related genes in intellectual disability (ID) that is only comparable with that reported for synapse-related genes.[2] This observation suggests that epigenetic regulation is a process of particular importance for the development and/or maintenance of cognitive abilities. In fact, most of these IDDs are caused by heterozygous mutations, suggesting that cognitive processes are very sensitive to gene dosage changes of epigenetic factors.[1] The list of epigenetic IDDs includes the ATR-X spectrum of disorders; immunodeficiency, centromeric instability, and facial dysmorphism (ICF) syndrome; Rett syndrome; Coffin–Lowry syndrome; and Rubinstein–Taybi syndrome (RSTS). See Table 6.1 and Kleefstra *et al.*[1] for a recent and exhaustive review.

Table 6.1 IDDs associated with mutations in genes directly involved in histone and DNA modification

Gene	OMIM[a]	Disease	Frequency[b]
Histone acetyltransferases			
CREBBP	180849 610543	Rubinstein–Taybi syndrome (RSTS) 1 Chromosome 16q13.3 deletion syndrome: severe RSTS form	1:100,000–125,000/ 1:300–500 of ID patients[c]
EP300	613684	Rubinstein–Taybi syndrome (RSTS) 2: mild RSTS	
KAT6B (MYST4/ MORF)	606170 603736	Genitopatellar syndrome (GTPTS) Ohdo syndrome, Say–Barber–Biesecker–Young–Simpson variant (SBBYSS)	1:1,000,000
Histone deacetylases			
HDAC4	600430	Brachydactyly mental retardation syndrome (BDMR)/2q37 deletion syndrome	Unknown
HDAC8	309585	Wilson–Turner X-linked mental retardation syndrome (WTS)	Unknown
Histone methyltransferase			
EHMT1	610253	Kleefstra syndrome: also mutations in other genes including epigenetic factors (MBD5, KTM2C/MLL3, SMARCB1, and NR1I3)	Unknown
EZH2	277590	Weaver syndrome (WCS): overlap with SOTOS1	Unknown
NSD1	117550	Sotos syndrome 1 (SOTOS1): overlap with WCS	1:5000–14,000
KMT2A (MLL)	605130	Wiedemann–Steiner syndrome (WDSTS)/hairy elbows, short stature, facial dysmorhism, and development delay	Unknown
KMT2D (MLL2)	147920	Kabuki syndrome 1 (KABUK1)	1:32,000
WHSC1	194190	Wolf–Hirschhorn syndrome (WHS): another mutation in D4S3327 marker region related to a mild form	1:20,000–1:50,000

Continued

Table 6.1 IDDs associated with mutations in genes directly involved in histone and DNA modification—cont'd

Gene	OMIM	Disease	Frequency
Histone demethylase			
KDM5C (JARID1C)	300534	Mental retardation, X-linked, syndromic, Claes–Jensen type (MRXSCJ)	Unknown
KDM6A	300867	Kabuki syndrome 2 (KABUK2)	Unknown
PHF8	300263	Mental retardation syndrome, X-linked, Siderius type (MRXSSD)	Unknown
Histone ubiquitination enzymes			
HUWE1	300706	Mental retardation, X-linked syndromic, Turner type (MRXST)	Unknown
	300705	Mental retardation, X-linked 17 (MRX17)	Unknown
UBE2A (RAD6A)	300860	Mental retardation, X-linked syndromic, Nascimento type (MRXSN)	Unknown
CUL4B	300354	Mental retardation, X-linked, syndromic 15 (Cabezas type) (MRXS15)	Unknown
DNA methyltransferase			
DNMT3B	242860	Immunodeficiency–centromeric instability–facial anomalies syndrome (ICF1): no apparent clinical differences with ICF2 caused by mutated ZBTB24	Unknown

[a]ID in the Online Mendelian Inheritance in Man database.
[b]According to Genetics Home Reference.
[c]Padfield *et al.*[3]
Note that genes encoding important enzymatic activities (such as ATPase helicase) that play a more indirect role in chromatin modification or remodeling processes are not included in this table.

1.1. Developmental and adult component of IDDs

The epigenetic modification of the chromatin, in the form of covalent modifications of the DNA or histones, plays a critical role in regulating most aspects of genome biology, from gene transcription to DNA replication and repair. These mechanisms are particularly relevant in metazoans because the epigenetic editing of the genome is the main mechanism underlying the specification and maintenance of different cellular phenotypes from a common DNA sequence. Thus, extensive chromatin remodeling occurs during early development. DNA methylation patterns change immediately after fertilization and during early-preimplantation development, along with changes in the profiles of specific histone modifications.[4] During the last decade, a wealth of studies has greatly contributed to decipher the epigenetic signals associated with stem cell fate decisions[5–7] and provided unprecedented view of the epigenetic changes involved in cellular differentiation and the maintenance of cell identity.[8–10] In particular, the relevance of epigenetic mechanisms in the maintenance of lineage specificity has been clearly demonstrated in experiments reprogramming adult cells into pluripotent stem cells in which a residual epigenetic signature characteristic of the donor cell restricted the subsequent differentiation to lineages related to the tissue of origin.[11] It is thereby not surprising that the mutation of genes encoding epigenetic factors causes severe developmental alterations even in hemizygous animals.

Epigenetic mechanisms also contribute to the activation and maintenance of specific differentiation programs in adult somatic cells.[4] In the context of IDDs, it would be particularly relevant to clarify whether the deficiency of these epigenetic factors causes ID by interfering with neurodevelopmental processes that later have consequences in brain function or by directly impairing cognitive processes in the adult brain. First, because the distinction between the developmental and the adult components of the disease has important implications in therapy (symptoms with a developmental origin can be indirectly compensated but not corrected, whereas deficits associated with altered functions in the adult brain seem better suited for direct pharmacological correction) and, second, because the existence of these disorders is frequently presented as one of the lines of evidence supporting a role for epigenetic processes in cognition. Confirming the prevalence of cognitive deficits independently of the developmental effects would greatly strengthen this evidence.

1.2. A network of epigenetic genes underlying IDDs

Intriguingly, IDDs originated by mutations in different genes encoding chromatin-modifying enzymes share features suggesting that common gene networks may be affected. For example, it has been recently found that Kabuki syndrome can be originated by mutations in the genes encoding the lysine methyltransferase KMT2D[12] or the lysine demethylase

KDM6A,[13] which have different catalytic activities and substrates. This discovery has led to the definition of two distinct genetic disorders with very similar clinical pictures known as Kabuki syndrome 1 (OMIM #147920) and Kabuki syndrome 2 (OMIM #300867), respectively.[14] Similarly, a combination of genetic studies in *Drosophila* and protein interaction analyses led to the identification of new genes associated with Kleefstra syndrome spectrum phenotypes.[15] The exploration of protein–protein interactions among ID-associated epigenetic factors has further unveiled the fine balance between agonistic and antagonistic epigenetic activities in play during cognitive development.[1] Further progress in the description of these syndromes and in the understanding of the precise sequence of epigenetic events underlying brain development is likely to reveal an even larger overlap between these pathologies.

Here, we will primarily focus on RSTS, which despite of its low prevalence is one of the most investigated IDDs with an epigenetic origin. In the next sections, we will summarize the current understanding of the etiology, pathophysiology, and treatment of this rare disease that emerge from studies in patients and animal models. As mentioned above, given the similarities between disorders originated by the deficiency of chromatin-modifying enzymes, this knowledge may also influence the research and therapy of other IDDs.

2. MOLECULAR GENETICS OF RSTS

2.1. Rubinstein–Taybi syndrome

RSTS is a complex autosomal dominant disease affecting 1 out of 100,000 newborns that is characterized by ID among other clinical symptoms. Rubinstein and Taybi first described this syndrome 50 years ago based on the distinct combination of delayed cognitive development and characteristic physical features.[16] RSTS patients exhibit skeletal abnormalities, such as defects in angulation, positioning, or duplication of phalanxes and thumb halluces that produce the characteristic broad thumbs and big toes that are used in diagnosis. Their facial appearance is also distinctive, with highly arched eyebrows, long eyelashes, highly arched palate, mild micrognathia, broad nasal bridge, beaked nose, and characteristic grimacing smile.[17–19] Neuroanatomical defects, including agenesis of the corpus callosum, cortical clefts, and subtler anatomical defects, have been also observed in the brain of some RSTS patients.

Psychological studies have shown that RSTS patients have short attention spans, poor motor coordination, and autistic features.[20] ID range from moderate to severe (IQ ranging from 80 to 25) and its severity may increase with age.[21] Worsening of behavioral symptoms over time has only been reported in about one-third of the patients, but this may be an underestimation because most RSTS cases are not under geneticists' surveillance in adults. Other neurological symptoms include seizures (28%) and abnormal electroencephalograms (~60%).[22] RSTS patients are also prone to develop tumors with a neural origin, such as neuroblastoma, medulloblastoma, and oligodendroglioma.[19,23] In addition to these skeletal and neurological symptoms, approximately a third of RSTS patients have heart malformations, such as atrial or ventricular septal defects and may present abnormalities in other tissues, such as the kidney and skin. In fact, complications from congenital heart disease and infections of the respiratory tract have been reported as the primary causes of morbidity and mortality in RSTS.

Although nowadays most RSTS patients survive infancy and reach adulthood, there are only a few documented cases of autosomal dominant transmission[17,24] being the majority of RSTS cases attributable to *de novo* mutations.[19]

2.2. Phenotype–gene relationships

The first gene associated with RSTS was *CREBBP* that encodes the cAMP response element-binding protein (CREB)-binding protein (CBP),[25] which receives this name because it was originally described as an interacting partner for the stimulus-regulated transcription factor (TF) CREB.[26] Ten years later, the mutations in a related gene, *CREBBP*'s paralogue *EP300*, were identified as an alternative cause of RSTS.[27] *EP300* encodes p300 that was initially described as the host factor interacting with the protein E1A from adenovirus type 5.[28,29] The function and structure of these proteins will be discussed in detail in another section.

The prevalence of mutations in *CREBBP* and *EP300* in RSTS patients is 50% and 8%, respectively.[30] According to this information, the OMIM database has recently created separated entries for two forms of the syndrome. The cases of RSTS associated with *CREBBP* mutations are now known as RSTS1 (OMIM #180849), while the cases linked to *EP300* mutations are referred to as RSTS2 (OMIM #613684). RSTS2 shows, in general, milder symptomatology than RSTS1 and lacks a clear distinct symptom for diagnosis like the broadening and angulation of thumbs and halluces characteristic of RSTS1, which raises to the possibility that some cases of RSTS2 go undetected.[30–34]

The mutations associated with RSTS are very heterogeneous. In the case of *CREBBP*, 160 different mutations have been identified so far (Leiden Open Variation Database v.2.0 Build 35), including deletions, insertions, rearrangements, and missense, nonsense, and splicing substitutions without any apparent hot spot in the open reading frame of *CREBBP*[35] (Fig. 6.1). In contrast, less than 30 patients have been described to date bearing mutations in *EP300*[30] (Fig. 6.2). The most extensive genotype/phenotype correlation study conducted in RSTS1 investigated a cohort of approximately 100 patients, including single amino acid changes, truncating and splice site mutations, and microdeletions that were distributed throughout the gene.[22] Despite this diversity, there are relatively few phenotypic differences between patients grouped by the type of mutation, other than a trend toward increased severity of ID and autistic features in patients with larger deletions. Thus, the mean age of diagnosis is much lower in patients

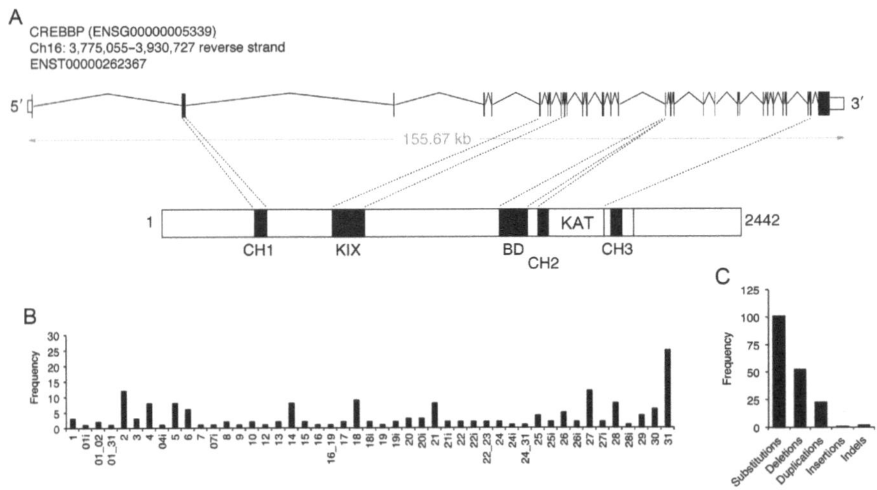

Figure 6.1 Genotype–phenotype relationships in RSTS1. (A) Schematic diagram of the *CREBBP* gene (top) and protein (bottom). Exons encoding for functional domains are indicated with dashed lines (CH, cysteine–histidine-rich; KIX, CREB-binding; BD, bromodomain; KAT, lysine acetyltransferase). (B) Distribution of 180 CREBBP variants reported so far, according to LOVD database (LOVD v.2.0 Build 35) across the 31 protein-coding exons. Insertions are indicated with lowercase "i"; deletions are indicated with underscore character "_." (C) Distribution of 180 cases of CREBBP mutation identified so far, according to LOVD database (LOVD v.2.0 Build 35) relative to the type of mutation (substitution, deletion, duplication, insertion, and indel).

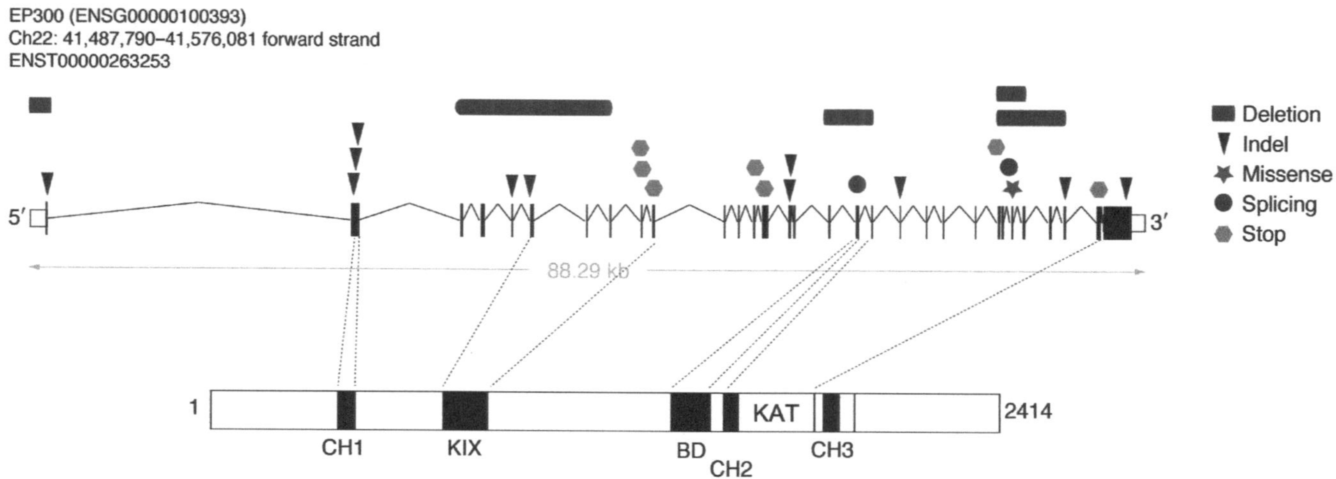

Figure 6.2 Genotype–phenotype relationships in RSTS2. Schematic diagram of the *EP300* gene (top) and protein (bottom). The 26 mutations identified so far are indicated with symbols depicting distinct types of mutations (deletion, indel, missense, splicing, and stop). Exons encoding for protein functional domains are indicated with dashed line (CH, cysteine–histidine-rich; KIX, CREB-binding; BD, bromodomain; KAT, lysine acetyltransferase). *See Ref. 30 for further detail.* (See the color plate.)

with large deletions.[36] Gene deletions range from several hundred kilobases to microdeletions and overall account for about 30% of RSTS cases. The most severe cases die during infancy, which is an otherwise rare event in RSTS, and define a different syndrome known as chromosome 16p13.3 deletion syndrome.[37] On the opposite site of the spectrum, some mutations in the *CREBBP* locus have been associated with "incomplete RSTS," a term introduced by Bartsch and colleagues to describe milder cases of the disease.[38]

Importantly, it should be noted that together the mutations into the *CREBBP* and *EP300* loci only account for less than two out of three cases diagnosed as RSTS. Although the remaining patients may contain mutations in regulatory sequences affecting these genes (note that genetic screens normally focus on the coding sequence), it is also possible that other still unidentified genes contribute to the development of RSTS. The identification of these additional genes would greatly broaden our understanding of RSTS etiology.

2.3. RSTS-related disorders

CREBBP mutations have been also associated with other pathologies beyond RSTS. As indicated above, the name chromosome 16p13.3 deletion syndrome (OMIM #610543) is used to denote the syndrome caused by deletions of chromosome 16p13.3 that go beyond the *CREBBP* locus and also affect the *DNASE1* and *TRAP1* genes, which encode the deoxyribonuclease I and the tumor necrosis factor receptor-associated protein 1, respectively. These cases are more severe than classic RSTS and normally show early death in infancy and life-threatening malformations.[37]

Intriguingly, gene duplications that are complementary to the deletion found in the most severe cases of RSTS cause a distinctive disorder referred to as chromosome 16p13.3 duplication syndrome (OMIM #613458). The patients diagnosed with this disorder show ID ranging from "mild" to "moderately delayed with speech problems," although some had low-normal intelligence. In addition, most patients show abnormal facial features and mild skeletal anomalies of the hands and feet, such as proximally implanted thumbs, long tapering fingers, and short toes.[39] Although these skeletal abnormalities do not exactly replicate the classical symptoms of RSTS, it is remarkable that both the hemideficiency and the duplication of *CREBBP* led to ID and anatomical problems affecting the same organs. This indicates that the relationship between CBP activity and developmental processes related to skeletal morphogenesis and cognitive abilities is defined by an inverted-U curve and any imbalance in gene dosage can provoke an abnormal development.

Outside the central nervous system, there is an enormous body of literature describing the role of CBP in cancer because human tumors frequently bear mutations in the *CREBBP* gene and translocations affecting this locus cause acute myeloid leukemia and mixed-lineage leukemia.[40–43] This topic is, however, out of the scope of this chapter.

In addition to pathologies in which CBP/p300 is directly affected, other IDDs are caused by histone-modifying enzymes that indirectly interplay with CBP/p300 activity. The most remarkable example is the Floating–Harbor syndrome (FLHS, OMIM #136140), a rare genetic disorder that has been recently associated with heterozygous mutation in the gene that encodes the SNF2-related CBP activator protein.[44] It is thereby tempting to speculate that both proteins participate in overlapping regulatory mechanisms. A number of phenotypic features are shared between RSTS and FLHS, including short stature, delayed bone age, delayed speech development, and typical facial features.[45] Only a few cases of FLHS have been reported so far in which cognitive outcomes range from "normal" to "significant ID" without any clear correlate with the nature or site of the originating mutation, suggesting that additional genetic modifiers and/or environmental factors might be involved.[44,46]

Interestingly, a recent study has linked CBP with a noncongenital neurodevelopmental disorder. The exposure to alcohol during pregnancy causes alterations in the development of the cerebellum that lead to fetal alcohol spectrum disorder (FASD), a condition that affects 1 in 100 children in the United States and currently is the leading cause of mental retardation in the United States. Guo and colleagues, working with a rat model for FASD, found that exposure to ethanol during late pregnancy reduces CBP levels in the cerebellum, which may contribute to the motor coordination and cerebellar-dependent learning task deficits that characterize FASD.[47]

2.4. Function of the KAT3 enzymes CBP and p300

CBP and p300 are highly homologous transcriptional coactivators with intrinsic lysine acetyltransferase (KAT) activity (Fig. 6.3). For this reason, they were renamed few years ago as KAT3A and KAT3B, respectively. They are the only two members of the KAT3 family. Both are large ubiquitously expressed nuclear proteins with an approximate molecular mass of

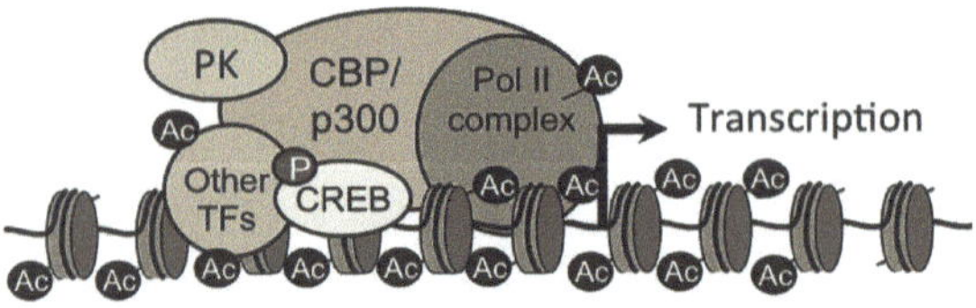

Figure 6.3 KAT3 proteins regulate transcription at different levels. CBP and p300 play a complex role transcription through different activities: (a) molecular bridge between TFs and the RNApolII complex, (b) molecular scaffold that brings a variety of enzymatic activities together to the promoter, (c) KAT that catalyzes the transfer of acetyl groups to lysine residues in the histones tails, and (d) KAT that catalyzes the transfer of acetyl groups to lysine residues to nonhistone substrates such as TFs or subunits of the RNApolII complex. TF, transcription factor; PK, protein kinase; Ac, acetyl group. (See the color plate.)

250 kDa that interact with numerous (>300) TFs of all major families.[48,49] Both proteins display widespread occupancy of transcriptional regulatory regions including many putative enhancers.[50–53] Also, both proteins share a similar structure with differentiated functional domains, including several cysteine-/histidine-rich regions (CH1 to CH3) involved in protein–protein interactions, the KIX domain that mediates interaction with phosphorylated CREB and other TFs, the KAT domain, a bromodomain (BD) that binds acetylated lysines, and two transactivation domains.[54] This modular structure makes possible that CBP and p300 function as molecular scaffold between DNA-binding TFs and the RNApol II complex.[48,55–57] Based on these interactions, both KAT3 enzymes are believed to play a critical role in transcriptional initiation. For example, in the case of the CREB-dependent gene expression, it is known that this TF is constitutively bound to specific sequences in many genes but remains inactive until its phosphorylation by stimulus-regulated kinases provokes the recruitment of KAT3 coactivators to the promoter triggering transcription.[26]

As indicated, in addition to the scaffolding function, both CBP and p300 have enzymatic activity as acetyltransferases of lysine residues. Although the role of this form of posttranslational modification (PTM) of proteins is still poorly understood, recent studies indicate that lysine acetylation is very ubiquitous and affects hundreds of proteins located in different cellular compartments and belonging to many important signaling pathways.[58–62] Opposing the action of KATs, lysine deacetylases (KDACs), also known as histone deacetylases (HDACs),[63] remove the acetyl group from lysines defining a dynamic balance of still uncertain function. The functional role of lysine acetylation in substrates of p300 and CBP has been only explored in relatively few instances. The better-known examples of lysine acetylation occur at the N-termini of the four nucleosome histones (H2A, H2B, H3, and H4), frequently referred to as the histone tails. These tails are unstructured and have a high intrinsic flexibility that enables their dynamic interaction with DNA. The addition of an acetyl group to positively charged lysine residues neutralizes their charge and loosens up the contact between histones and DNA. This conformational change is thought to cause the relaxation of the chromatin and facilitate the access of TFs and the basal transcriptional machinery to specific DNA sequences. The balance between histone acetylation and deacetylation is, therefore, thought to contribute to define global patterns of transcriptional competence[64] and might also contribute to the local regulation of transcription initiation and elongation at specific loci, although the later role is still under debate.[65,66] Lysine acetylation has been also associated with transcription through histone-independent mechanisms. Thus, the acetylation of specific lysine residues in the TFs p53 and NF-κB contributes to regulate their transactivation activity,[67,68] and the acetylation of the regulatory subunit of the RNA polymerase II RPB1 plays a critical role in the transcriptional activation by growth factors.[69]

As we will see in the next section, despite sequence homology (>60%) and many common interaction partners,[70] CBP and p300 proteins have distinct and nonredundant functions in the brain and elsewhere.[49,56]

3. LESSONS FROM MOUSE MODELS

Basic research on genetically modified mouse strains designed to reproduce the genetic alterations underlying congenital disorders represents today the most valuable approach to progress in the understanding and cure of these conditions. In particular, in IDDs, given their syndromic nature and the affection of superior cognitive functions, mouse models are more suitable than alternative systems, such as cultured cells or *in silico* models, to investigate the underlying mechanisms and assess possible therapies. In addition, second-generation mouse genetics tools, such as inducible conditional knockouts (cKOs) and regulatable and tissue-specific transgenics, allow the dissection of distinct anatomical and temporal components of the disorder.

RSTS represents a prime example of the utility of mouse models because most of the progress in our understanding of this disorder comes from the generation and characterization of mouse strains that reproduce the genetic alteration mutations found in patients. In the next sections, we first briefly describe the different mouse models currently available for investigating

RSTS to later review the insight into the role of CBP and p300 in normal brain function and disease achieved through the characterization of these models.

3.1. Different models address different questions

Different CBP- and p300-deficient mouse strains have been generated to investigate specific aspects of RSTS pathology and elucidate the function of these proteins in the nervous system (Fig. 6.4).

a. *Null mutants*: Conventional heterozygous knockout mice for either *Crebbp* (*cbp*$^{+/-}$) or *Ep300* (*p300*$^{+/-}$) are the most appropriate models to reproduce the complex spectra of pathological features described in RSTS because they bear genetic disruptions that are equivalent to the hemizygous mutations found in patients. These mice allow the exploration of cognitive, skeletal, or hematologic defects (among others) that

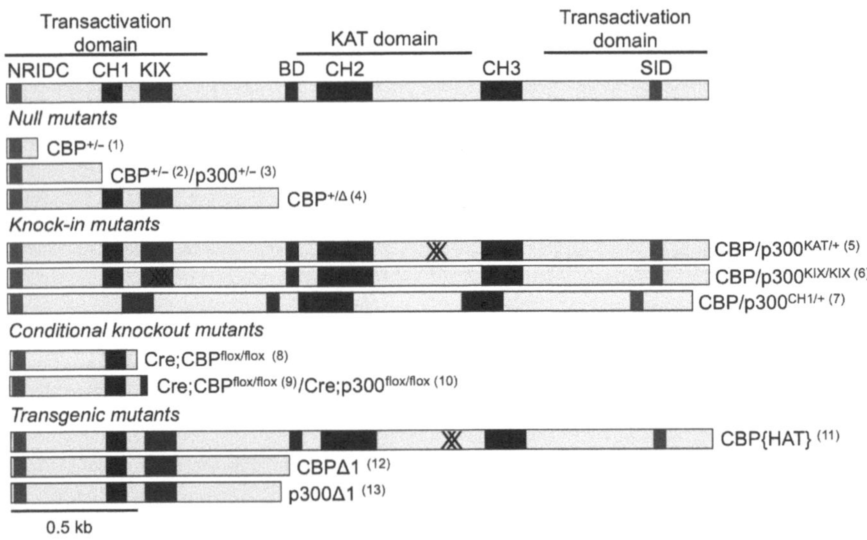

Figure 6.4 Linear representation of wild-type CBP and p300 and mutant variants expressed in mouse models of RSTS. The location of some important domains and sites in CBP structure is labeled; these include a lysine acetyltransferase (KAT) domain, a bromodomain (BD), two transactivation domains, three cysteine/histidine-rich regions (CH1 to CH3), and multiple specific interaction domains for different TFs, such as the CREB-interacting domain KID and the nuclear receptor (NR)-interacting domain at the N-terminus of the protein. (1) Tanaka *et al.*[71]; (2) Kung *et al.*[72]; (3) Yao *et al.*[73]; (4) Oike *et al.*[74]; (5) Roth *et al.*[75]; (6) Kasper *et al.*[76]; (7) Kasper *et al.*[77]; (8) Zhang *et al.*[78]; (9) Kang-Decker *et al.*[79]; (10) Kasper *et al.*[49]; (11) Korzus *et al.*[80]; (12) Wood *et al.*[81]; (13) Oliveira *et al.*[82] (See the color plate.)

are reminiscent of disease symptomatology, as well as the assessment of possible therapies. Three independent knockout strains have been generated in which the *Crebbp* gene has been targeted. The first of these strains was generated through the removal of the exons corresponding to aas 29–265, which is expected to completely eliminate the expression of the protein.[71] The second was generated by the insertion of a gene-trapping vector in introns 16–17 that left intact the first half of the protein, including the KIX domain, but disrupted the bromodomain and subsequent domains.[74] In the third strain, the gene-targeting strategy eliminated the exons encoding aas 340–443 and is expected to produce a short protein that includes the first Cys/His domain (CH1).[72] The last strategy was also used to generate the only knockout strain currently available for *Ep300*. In that case, the truncation of the protein after CH1 was achieved by removing the exons corresponding to aas 302–509.[73] Although in all these studies the aim was to produce a truncated nonfunctional protein, in the case of the mutant generated by Oike and colleagues, a dominant negative effect has been reported.[74]

b. *Knock-in mutants*: Restricting the mutation to specific residues is a useful approach to gain insight into the molecular mechanisms underlying the disease by assessing the involvement of specific protein domains. Three domains of CBP/p300 have been targeted using knock-in procedures, namely, the KAT, KIX, and CH1 domains. In the first case, inactivating point mutations were introduced in the KAT domain of either CBP (residues W1503 and Y1504) or p300 (W1466 and Y1467) to study the involvement of these KAT activities in organ formation.[75,83] In the case of the KIX domain, three point mutations affecting Y650, A654, and Y658 in either CBP or p300 diminished the interaction of these coactivators with specific partners, such as the TFs c-Myb and CREB.[76] Finally, the role of the CH1 domain in transcriptional transactivation was explored by deleting the exons encoding aas 342–393 in CBP or 329–379 in p300.[77,84]

c. *cKOs*: The investigation of the precise role of KAT3 proteins in normal and diseased brain is hampered by the prenatal lethality of null mutations in homozygosis. This situation can be circumvented through the use of cKO mice in which the expression of the mutation can be restricted spatially and/or temporally. For example, neuronal-restricted ablation can be achieved through the crossing of *floxed* mice (animals in which *loxP* sites have been introduced flanking specific exons) with transgenic mice that express the Cre recombinase in neurons; this would result in the elimination of the flanked exons in neuronal tissue and the consequent loss of expression or production of a truncated protein. The use of inducible recombinases, which trigger gene ablation at the time decided by the experimenter, enables an even more precise dissection of the deficits.[85] Such strains allow the dissection of the developmental and adult components of the disorder and a more precise examination of the role of KAT3 proteins in cognitive function by eliminating the confounding effects of growth retardation, skeletal abnormalities, reduced weight, and cardiac deficits observed in conventional KO mice. Two strains of floxed *Crebbp* mutants have been generated in which loxP sites have been introduced flanking exon 7[78] or 9,[79] respectively. In the case of *Ep300*, the only available cKO strain was generated by flanking exon 9.[49] These strains have been used to target different cell types and tissues, including thymocytes,[79] renin-expressing cells,[86] and postnatal forebrain neurons.[87,88] Importantly, cKO strains can be also used in combination with recombinant viruses expressing Cre recombinase to target highly specific cell subpopulations, such as neurons in the CA1 and DG subfields of the hippocampus[89] or in the nucleus accumbens.[90]

d. *Transgenic mutants*: The expression of dominant negatives mutants that interfere with the normal function of endogenous KAT3 proteins of interest represents a powerful alternative to gene targeting. Transgenic approaches based on the use of tissue or cell-specific promoters can help to determine the contribution of specific brain areas or subcellular types to particular aspects of the pathology. Furthermore, some transgenic approaches also provide temporal control of transgene expression that can be used to dissect the developmental and adult components of the disease and to explore the reversibility of the phenotype, which is not possible in cKO. Thus, the first transgenic model of RSTS, referred to as CBP{HAT−} mice, expressed under the control of the CaMKIIα-tTA/tetO system a mutated CBP bearing two point mutations (Y1540/F1541) in the KAT domain that completely abolished its enzymatic activity but preserved other protein–protein interactions.[80] In this system, the transgene is specifically expressed in postnatal forebrain neurons, and expression can be turned on or off by the removal or addition of the drug doxycycline from the diet.[91] The same transgenic approach was later used to express a truncated p300 (aas 1–1031) lacking the KAT domain and the C-terminal half of the protein.[82] The same group used a simpler transgenic strategy to express a similarly truncated CBP (aas 1–1083) under the direct control of the CaMKIIα promoter.[81]

3.2. KAT3 proteins in neurodevelopment

The KAT3 proteins CBP and p300 play a key role during development, controlling the proliferation and differentiation of different cell lineages, including those within the nervous system.[92] In mammals, both proteins are first required for neural tube closure at the three-layer embryonic stage. Homozygous mice bearing *cbp*- or *p300*-null alleles die early during development, between E9 and E12.5.[73,74,93] Death may occur by exencephaly[72,73,93] and abnormal blood vessel formation[74,93] and heart development.[73] Strikingly, the double heterozygous mutation for CBP and p300 is also lethal,[73] indicating that these two proteins play redundant roles during early development and that a certain total level of KAT3 activity is essential at this stage. Of note, disruption of neither one of the proteins was compensated by the upregulation of the other at later stages.[87,88,94] The functional overlap is lower when hemizygotic mutations for other KAT proteins are considered.[95] Thus, $GCN^{+/-}/p300^{+/-}$ showed limited embryonic lethality (~25% of animals) and neither $PCAF^{+/-}/p300^{+/-}$ nor $PCAF^{-/-}/p300^{+/-}$ exhibited early death. Interestingly, knock-in mutations affecting either the KAT or the CH1 domains caused a stronger phenotype that the complete loss of the protein because these mutants exhibit embryonic or neonatal lethality even in heterozygosis.[77,83] These results, consistent with early experiments *in vitro*,[96] suggest a dominant effect of mutated proteins that contain a fully functional KIX domain. Of note, both viability and penetrance of RSTS-like phenotypes are dependent on the genetic background in which mice are inbred,[71,73,94] suggesting that multiple loci interact with KAT3 genes during embryonic development.

At the time when the neural plate is folding to generate the neural tube, both CBP and p300 are highly expressed decaying their expression afterward.[97,98] Despite the decrease, both proteins are still involved in the differentiation of neurons and glial cells from cortical precursors at later stages.[99] For example, it has recently been shown that interneuron development is transiently compromised in *Cbp*-deficient mice,[100] a finding that could underlie the high incidence (>25%) of seizure disorder development among RSTS patients.[22,101] Consistent with the developmental role of CBP and p300 in mammals, early work in *Drosophila* has demonstrated that mutations in dCBP (also known as *Nejire*) result in impaired mesoderm differentiation.[102] More recently, evidence from genome-wide binding profiles of 40 TFs indicates that dCBP shows preferential co-occupancy of sites bound by TFs that regulate dorsoventral patterning.[103] Studies focusing on latter developmental states have shown that dCBP activity is also critical for neuronal differentiation in the fly retina[104] and affects glial-cell development.[105]

The comparison of CBP- and p300-deficient mice has revealed important differences in the role of these two proteins during development outside the brain. For example, embryos with disrupted p300 KAT activity show pronounced organogenesis defects affecting the skeletal muscle,[75] heart and coronary vasculature, lung epithelium, and small intestine[83] that are not observed in equivalent CBP mutants, In contrast, multilineage defects in hematopoiesis and increased incidence of hematologic malignancies are more prominent in CBP mutants[49,72,79] although p300 is also essential to regulate other aspects of hematopoietic differentiation.[77] These functional divergences may be better explained by differential protein interaction and substrate specificity[56] but, unfortunately, a precise catalog of such interactions *in vivo* is still lacking.

3.3. Modeling syndromic manifestations

Heterozygous *Crebbp* and *Ep300* mice ($cbp^{+/-}$ and $p300^{+/-}$) are viable and fertile but exhibit craniofacial dysmorphia (prominent forehead, blunt nose, and large anterior fontanel) and other skeletal abnormalities associated with RSTS, such as abnormal rib cage and axial bones.[71,74,94] The characteristic broad thumbs and toes used for RSTS diagnosis are not seemingly reproduced in mice. Interestingly, 3-phosphoinositide-dependent protein kinase-1 (*Pdpk1*) knockout mice show skeletal abnormalities similar to those described in $cbp^{+/-}$ mice,[72] and the overexpression of bone morphogenetic protein 2/7 (BMP2/7) leads to amelioration of the phenotype in both mutants,[106] suggesting that the two proteins are part of the same regulatory pathway during skeletogenesis and osteoblast differentiation.

In addition to skeletal defects, hemizygous mutants also show growth retardation,[71,72,74,94] increased insulin sensitivity and glucose tolerance,[107] and defects in hematopoiesis with a tendency to hematologic malignancies[72] that is consistent with the ascribed role of KAT3 proteins as tumor suppressors[108] and with the increased cancer risk observed in RSTS patients.[109] Although the different hemizygous mutants have never been compared in parallel, the strain generated by Oike and colleagues, which is predicted to express a truncated CBP protein after the KAT domain, seems to exhibit more severe phenotype than null hemizygous mice, including cranioskeletal abnormalities, poor locomotor activity, and cardiac defects. As already described for KAT knock-in mutants, this result is interpreted in terms of a dominant negative effect of KAT-dead mutant proteins.[74] In contrast to KAT knock-in mice, homozygous $CBP^{KIX/KIX}$ and $p300^{KIX/KIX}$ mice are viable, although in the case of p300 some deceases were observed around the age of 3 weeks and surviving animals tended to be smaller and showed hematopoietic defects. Heterozygous knock-in mutants for the CH1 domain are also viable and only show a slight deviation from Mendelian inheritance in the case of CBP, although some RSTS characteristics such as craniofacial abnormalities and reduced body size were observed in combined heterozygous for both proteins ($CBP^{+/\Delta CH1}$ and $p300^{+/\Delta CH1}$).[77] All the investigated strains of floxed mice ($CBP^{flox/flox}$ or $p300^{flox/flox}$) are undistinguishable from wild-type littermates in the absence of cre recombinase.

3.4. Cognitive impairments

In an attempt to model ID, the different mouse models of RSTS have been examined in a variety of tasks assessing different forms of memory. Conversely to other syndromic features, cognitive deficits have been observed in all the examined strains including hemizygous mice, cKOs, transgenic strains expressing dominant negative variants, and mice with focal ablation achieved through viral transduction. These experiments have revealed memory defects in aversive associative memory (fear-conditioning and step-through passive avoidance tasks), episodic memory (novel object recognition task), and spatial memory (Morris water maze task), among other behavioral traits related to ID.[110] Overall, the studies underscored the relevance of CBP and p300 functions in neuronal plasticity-related processes, including learning and memory[54] and drug addiction.[90,111] However, they have also revealed important differences concerning the magnitude of the deficits and the specific behavioral trait affected in the different strains. We refer the readers to Table 6.1 in our recent review on the role of KAT3 proteins as therapeutic targets in cognitive and neurodegenerative diseases for a detailed comparison of the phenotype observed in the different mutant strains.[54]

Notably, all the studies conducted so far have reported deficits in a specific form of episodic memory known as novel object recognition,[87–89] suggesting a specific role for CBP in the consolidation of this form of memory. Surprisingly, the comparison of results obtained from hemizygous mice and cKO suggests that the complete loss of CBP does not worsen the deficits associated with hemideficiency. On the contrary, phenotypes were frequently stronger in hemizygous mice underscoring the contribution of developmental and/or nonneuronal effects to the cognitive impairments.[88,112] In hemizygous strains, cognitive impairment is often intermixed with motor deficits, such as reduced swim velocity in the water maze and motor coordination impairment in the accelerating rotarod task,[94,112,113] introducing a confounding factor that should be considered in the interpretation of memory experiments. Of note, RSTS patients experience difficulties in planning and executing locomotor skills[114]; therefore, the detection of motor impairments in CBP-deficient mice contributes to the modeling of the disease.

The comparison of studies investigating cognitive impairment in CBP cKO mice has revealed important differences between labs and strains. For example, a recent study on cKO mice in which CBP was completely inactivated in excitatory neurons of the postnatal forebrain described deficits in both long-term memory (LTM) and short-term memory (STM),[87] suggesting that the role of CBP in memory storage may also encompass memory encoding. However, this effect was not observed in similar studies conducted in our laboratory,[88] nor in experiments with focal CBP ablation.[89]

Another interesting conclusion of the comparison across studies is that p300-deficient mice show weaker cognitive defects than CBP-deficient mice.[94] For instance, the analyses of $p300^{+/-}$ mice revealed mild impairments in the water maze along with minor motor defects but no impairment in the fear-conditioning or object recognition memory tasks nor long-term potentiation (LTP).[94] These results are consistent with the milder ID observed in patients bearing mutations at *EP300* compared to *CREBBP*.[31,33,34] Like in the case of CBP, the expression of a truncated p300 protein may cause more severe deficits than hemizygosis. Thus, transgenic mice expressing a truncated *p300* protein ($p300^{\Delta 1}$) in the hippocampus, amygdala, cortex, and cerebellum exhibited memory deficits in contextual fear-conditioning and object recognition tasks.[82] Similar memory deficits have been also reported in cKO mice with a complete ablation of p300 restricted to CA1 and the cortex.[115]

In some of studies, cognitive deficits were investigated in parallel with electrophysiological recordings. LTP in CA3–CA1 synapses in the hippocampus is considered a cellular correlate of explicit memory formation. In particular, the late phase of LTP (L-LTP) is, like LTM, dependent on *de novo* transcription. Consistent with this view, L-LTP was found impaired in the hippocampus of different CBP mutants, including the heterozygous null mutant,[112] the transgenic mouse expressing a truncated CBP variant,[81] and the focal cKO,[89] whereas the early phase of LTP, which is independent of *de novo* transcription, was unaffected. Importantly, basal synaptic transmission was in all the cases normal.

3.5. Adult neurogenesis

We have recently described a new phenotype associated with CBP deficiency: deficits in experience-induced adult neurogenesis.[116] In the subventricular zone of the lateral ventricles and the subgranular zone of the hippocampal dentate gyrus of the mammalian brain, adult neural stem cells proliferate and differentiate to newborn neurons throughout adulthood.[117–119] These newly generated neurons functionally integrate into circuits and contribute to cognitive function and plasticity.[120–122] In particular, hippocampal neurogenesis has been involved in specific aspects of memory, such as spatial pattern discrimination and memory reorganization during system consolidation.[120,123,124] Importantly, adult neurogenesis is dynamically regulated by the cooperation of intrinsic state-specific cell differentiation cues and extrinsic neural niche signals,[125] and, similar to the situation during embryonic development, epigenetic mechanisms are thought to contribute this regulation.[126,127]

Studies in CBP-deficient mice indicate that this transcriptional coactivator is neglectable for steady-state neurogenesis in the adult hippocampus.[116,128] However, experiments in both conventional and conditional KO mice revealed that CBP becomes again necessary for modulating the rate of neurogenesis in response to changes in the activity of the hippocampal network, such as those driven by environmental enrichment (EE) or epileptic activity.[116] Importantly, the deficit in environmentally induced neurogenesis coincided with a failure to express the improvement in spatial memory that would be normally associated with EE. Of note, adult hippocampal neurogenesis was not affected in p300 heterozygous,[116] further confirming the different relevance of these two proteins in brain processes.

Both main functions of CBP (transcriptional coactivator and KAT) can contribute to this defect. On the one hand, different TFs involved in adult neurogenesis have CBP as coactivator.[129,130] On the other hand, a number of pharmacological and genetic experiments indicate that lysine acetylation plays an important role in neurogenesis.[131] For example, TTK21, a small-molecule drug that stimulates the KAT activity of CBP/p300, promotes maturation of newborn neurons in the hippocampal dentate gyrus and increases spatial memory duration,[132] whereas the treatment with HDAC inhibitors (HDACis) reduces neural cell proliferation and promotes neuronal differentiation *in vitro*.[133–136]

3.6. Molecular etiology of RSTS

Biochemical and molecular biology experiments in animal models of RSTS suggest that impaired KAT3 function, affecting both transcriptional and epigenetic regulations, may contribute to the etiology of the phenotypes described in the previous sections.

Among the numerous proteins that interact with CBP and p300, there are a number of TFs involved in cognitive processes, such as CREB and FOS.[137] In particular, given the prominent role of CREB as part of the molecular switch that transform STM into LTM,[138,139] the impairment of CREB-dependent transcription can critically contribute to the cognitive deficits observed in RSTS. Supporting this view, CRE-dependent transcription is impaired in cell lines from RSTS patients bearing mutations in the KAT domain of CBP[140] and in neurons of transgenic mice expressing a dominant negative CBP variant.[80,141] Furthermore, enhancement of CREB activity ameliorates synaptic plasticity deficits in $cbp^{+/-}$ mice.[112] However, it should be noted that experiments in CBP heterozygous and forebrain-specific cKO mice indicate that the reduction or loss of CBP has a modest impact in activity-driven transcription.[88,112] Intriguingly, whereas the hippocampal induction of immediate early genes in response to novelty was only moderately affected by CBP deficiency,[88] the transcriptional program induced in response to long-term EE was clearly impaired,[116] suggesting that CBP may be more important for the establishment of enduring changes in gene expression than for temporal transcription regulation. The requirement for p300 in CREB-dependent transcription is also not clear. Although p300 can activate CRE-dependent transcription,[142] experiments in fibroblasts from $p300^{-/-}$-null mice stimulated with the catalytic subunit of PKA have revealed normal activation of a CRE-luciferase reporter.[73] Gene expression profiling in the hippocampus of p300 heterozygous mice did not reveal either a specific impairment of CREB-dependent transcription.[94]

The research on RSTS mouse models has shown that impaired neuronal histone acetylation is the second hallmark of the disease that might play a role in the etiology of the symptoms according to the extensive correlative evidence supporting a role for histone acetylation in mnemonic processes.[143,144] Notably, experiments in lymphoblastoid cell lines have revealed similar histone acetylation deficits in RSTS patients.[145] However, we should remember that these KATs also regulate the acetylation of many other proteins that are likely affected by the reduced KAT activity and can contribute to the observed phenotypes. Moreover, the acetylation of lysines at the histone tails is not the only histone PTM described as affected by CBP hypofunction. The monoallelic deletion of *CREBBP* leads to the induction of the ERG-associated protein with SET domain (SETDB1, *aka* ESET), a histone-specific trimethyltransferase that increases the trimethylation of histone H3 at K9 (H3K9me3) and ultimately leads to the condensation of pericentromeric heterochromatin structure in neurons and striatal neuron atrophy and dysfunction.[146]

4. FROM MOUSE MODELS TO PATIENTS: THERAPY PROSPECTS

Different forms of therapy, based on drug treatment, gene manipulation, or environmental stimulation, have been assessed in mouse models of RSTS revealing their potential efficacy for treating cognitive impairment. However, to the better of our knowledge, none of them have been examined through clinical trial in patients.

4.1. HDAC inhibitors

As already discussed, the action of KATs is antagonized by HDACs; hence, the deficiency of KAT activity can be compensated through HDAC inhibition. Four classes of HDACs (I, II, III, and IV) have been defined according to sequence homology and domain organization. Inhibition of HDAC activity has demonstrated notable preclinical potential to enhance memory and ameliorate cognitive deficits in different models of disease including RSTS.[54,147,148] Clinically relevant HDACis belong to different chemical families: hydroxamic acids (e.g., suberoylanilide hydroxamic acid (SAHA) and trichostatin A (TSA)), benzamides (e.g., MS-275), and carboxylic acids (e.g., butyrate and valproate).[149] Some compounds act as pan-HDACi, such as SAHA and TSA, whereas others display narrower substrate specificity, such as tubacin. In particular, different inhibitors of HDAC classes I and II have been shown to favor memory formation and stability in some hippocampal-dependent tasks including contextual fear conditioning and novel object recognition,[141,150–157] although not all studies have reproduced these findings.[158] HDACis also facilitate the induction of

L-LTP evoked at the CA3–CA1 synapse[112,152,157,159] and increase spine density.[159,160]

Regarding the situation in mice with deficient CBP activity, seminal experiments in the Kandel's lab showed that the HDACi SAHA restored the aforementioned deficits in hippocampal L-LTP and long-term fear memory observed in the null allele mouse model for RTS.[112] Similar experiments in CBP{HAT−} transgenic mice also revealed an amelioration of memory deficits by TSA.[80] Later studies have shown that the impairment in object recognition reported in $CBP^{KIX/KIX}$ mice[161] can be also rescued by HDAC inhibition[155] although lysine acetylation was, in principle, not affected in those mutants. Moreover, additional studies in $CBP^{KIX/KIX}$ mice have revealed that the beneficial effects of HDACi in synaptic plasticity and hippocampal memory require CBP.[156,162] In contrast, TSA failed to modulate learning and memory in mice showing complete loss of CBP function in forebrain,[87] and systemic administration of the HDACi sodium butyrate did not rescue memory deficits in mice with focal deletion of CBP in the CA1 area of the hippocampus.[89] Together, these results pinpoint to HDACis as promising therapeutic drugs to treat RSTS and other conditions in which CBP function is compromised. However, variables, such as time, dose, and via of administration, should be standardized in order to clarify some inconsistencies between the studies conducted at different labs.

Several HDACis are approved drugs to treat some forms of leukemia and more are being currently assessed in clinical trials[163] whose success may encourage the examination of the same drugs in the treatment of brain disorders. Although the experiments in mouse models of RSTS are certainly encouraging, we should remember that the mechanism of action of these compounds in the brain remains largely unknown. It has been proposed that the transcriptional stimulation of plasticity and prosurvival genes via KAT-mediated increase in histone acetylation may be responsible for the reported beneficial effects. However, as suggested in a recent study by our group,[164] it is possible that these effects are partially, or even totally, independent of histone acetylation. Other substrates of HDACs and KATs, such as TFs, could mediate the highly restricted transcriptional effects downstream of HDAC inhibition.

4.2. PDE inhibitors

The discovery of the phosphodiesterase (PDE) inhibitors (PDEis) was concomitant to the identification of PDEs, the enzymes that catalyze the hydrolysis of the phosphodiester bond in the second messenger cAMP and other cyclic nucleotides.[165] In the early 1970s, the PDE4 inhibitor rolipram was found to be effective to treat depression, but its side effects and the success of alternative approaches, such as serotonin-selective reuptake inhibitors (SSRIs), slowed down its clinical development.[166] However, the interest on PDEis was later renewed for treating memory disorders. Seminal studies in rodents showed that the increase of intracellular levels of cAMP following rolipram administration was associated with facilitated induction of L-LTP at the CA3–CA1 hippocampal synapses and enhanced memory in a variety of learning paradigms.[167–169] In addition to the potentiation of synaptic strength, PDEis might also enhance cognition through other mechanisms, such as promoting CREB-mediated neurogenesis.[170]

PDEis have been found to exert beneficial effects in mouse models of RSTS. Thus, a novel PDE4 inhibitor known as HT0712 improved object recognition memory in mice carrying a truncated *Crebbp* allele in a dose-dependent manner.[171] Furthermore, both rolipram and overexpression of a constitutively active form of CREB were shown to facilitate L-LTP induction in *cbp*$^{+/-}$ mice.[112] Interestingly, phase 2a clinical trials in subjects with age-associated memory impairment indicate that PDEis may improve the performance a word list memory task (http://www.dartneuroscience.com/ClinicalTrials.php). If these results are confirmed, future studies could also explore the potential of PDEis to treat cognitive deficits in RSTS patients.

4.3. Environmental enrichment and occupational therapies

Occupational therapy provides to patients affected by neurodevelopmental, neuropsychiatric, and neurodegenerative diseases specific activities and richer life experiences aimed to prevent cognitive decline and promote independent function in all aspects of daily life. Mouse models represent a valuable mean to assess the efficacy of these therapies in specific pathological conditions. To date, a wealth of data have confirmed the beneficial effects of sensory, cognitive, and motor stimulation in different animal models of brain disorders including Alzheimer's disease (AD), Huntington's disease (HD), fragile X syndrome, and Down syndrome.[172] Our team has recently shown that the exposure to EE also ameliorates cognitive deficits in mouse models of RSTS. Notably, EE caused a remarkable improvement in motor skill learning and contextual fear memory in *cbp*$^{+/-}$ mice, and these beneficial effects were associated with EE-mediated increase in synaptic density in the hippocampus.[116] These results indicate that occupational therapies may also ameliorate cognitive deficits in RSTS.

4.4. Adult neurogenesis as therapeutic target

Intriguingly, the study described above revealed that some aspects of neuroadaptation to EE, such as enhanced neurogenesis and the associated improvement in spatial pattern discrimination, were defective in *cbp*$^{+/-}$ mice.[116] This discovery uncovered an alternative front to tackle ID in RSTS: enhancing adult neurogenesis. To date, few therapeutic strategies have been shown to be effective in promoting adult neurogenesis in animal models. These include voluntary exercise and antidepressants such as SSRIs and norepinephrine-selective reuptake inhibitors.[173–175] A recent study has also shown that metformin, a compound widely used to treat type II diabetes and other metabolic syndromes, promotes neurogenesis *in vivo* in an aPKC–CBP pathway-dependent manner and enhances spatial memory formation.[128] On the other hand, TTK21, a small molecule that activates KATs CBP/p300, has also been recently shown to promote adult neurogenesis and to extend the duration of spatial memory in mice,[132] whereas the HDACi TSA completely rescued the decreased neurogenesis observed upon differentiation of CBP-deficient cortical precursors and medial ganglionic eminence cells.[100,128] We do not know yet whether the deficit in activity-dependent adult neurogenesis reported in mice also occurs in RSTS patients contributing to cognitive impairment. If this turns to be the case, RSTS patients might eventually benefit from therapeutic strategies aimed to modulate adult neurogenesis.

5. STANDING QUESTIONS AND CONTROVERSIES IN RSTS RESEARCH

Despite the notorious progress in our understanding of RSTS etiology and CBP/p300 function in brain development and function summarized in the previous sections, there are several outstanding questions related to RSTS whose answers have implications well beyond this rare disorder.

5.1. RSTS, a link between epigenetics and memory?

The discovery that IDD, such as RSTS, is caused by epigenetic enzymes together with the finding that different forms of epigenetic labeling of the chromatin are dynamically regulated during learning and memory processes has generated great interest on the role of epigenetic mechanisms in cognition.[176] Thus, several studies during the last decade have suggested that experience-driven histone acetylation changes along with other epigenetic mechanisms might allow neurons to customize their gene expression patterns according to experience, widening the computational and storage capabilities of neural networks.[143,144,177–181] In this conceptual frame, the mouse strains used to model RSTS also represent valuable loss-of-function models for understanding the role of KAT3 proteins and histone acetylation in the developing and adult brain. The experiments with conventional and conditional KO mice indicate that the impact of eliminating specific KAT activities during adulthood is relatively modest when compared to early developmental stages because the animals are viable and do not present any major pathology.[88,182,183] This view is consistent with the notion that epigenetic marks are critical during cell differentiation and lineage specification in multicellular organisms, whereas they would play a secondary, although still important, role related to the maintenance of active/silent gene states in differentiated cells.

Regarding a mnemonic role of histone acetylation, we should note that although extensive correlative evidence supports a role for histone acetylation in neuronal plasticity, only some cognitive deficits have been consistently reproduced across multiple CBP/p300-deficient mouse strains and labs. Further research should clarify the differences between studies and clearly determine the specific impact of the ablation of CBP in fully differentiated neurons in learning and memory.

5.2. Do KAT3 deficiency and histone hypoacetylation cause neurodegeneration?

A number of studies have suggested that neurodegeneration in conditions such as AD and HD could be a consequence of neuronal histone deacetylation and selective downregulation of genes involved in neuroplasticity and survival.[160,184–188] Moreover, in the case of HD, transcriptional dysregulation has been specifically linked to the aberrant interaction of CBP with mutant huntingtin and histone hypoacetylation.[189] However, consistently with the absence of neurodegeneration in RSTS patients, the results in cKO mice indicate that the complete neuronal loss of CBP is not associated with neurodegeneration and the concomitant dramatic reduction of histone acetylation does not impair neuronal viability.[87,88] We should, however, note that CBP and p300 deficiencies do not equally affect the acetylation levels of all histones and the residual marks may be sufficient to sustain neuronal survival. It is also possible that histone hypoacetylation only became detrimental under cellular stress conditions. Again, further research should clarify the postulate relationship between KAT/HDAC unbalance and neuropathology determining if it is causative or consequential.

5.3. What is the relationship between transcription and CBP-dependent lysine acetylation?

The relationship between specific PTM of histones, including lysine acetylation, and transcription, is a major outstanding question in the epigenetic field.[65,190–192] Although the acetylation of lysine residues in the histone tails is strongly associated with active transcription, we should however note that this notion is largely based on correlative evidence and a causal connection between acetylation marks and gene activation is still lacking. The mouse strains used to model RSTS are also valuable loss-of-function models to investigate the role of KAT3 proteins in transcription. Thus, gene profiling experiments have consistently demonstrated that the loss of these proteins only caused a limited number of transcriptional changes, which were in some cases very modest[87,88,94,116] Moreover, some transcriptional events of particular relevance in plasticity, such as activity-driven gene induction, seem largely spared in KAT3-deficient mice.[88,112] These gene-specific effects contrast with the widespread histone hypoacetylation suggested by Western blot and immunohistochemistry experiments.[88] In agreement with these observations, experiments in mouse embryonic fibroblasts (MEFs) derived from the KIX knock-in mice revealed a significant impact in only a subset of CREB-dependent genes.[193] Moreover, experiments in MEFs from double knockout mice for CBP and p300 showed almost complete loss of histone H3K18 and H3K27 acetylation but limited transcriptional effects.[194] Intriguingly, some of these transcriptional deficits were rescued by overexpressing an alternative coactivator of CREB, the CREB-regulated transcription coactivator 2 (CRTC2), that lacks the KAT activity of CBP.[183] Future genomic experiments exploring the correlation between specific histone acetylation profiles and transcriptional deficits in the context of RSTS should importantly contribute to clarify the relationship between impaired lysine acetylation and transcription.

6. CONCLUDING REMARKS

The progress in human genetics and animal model research has produced a rapid and notorious advance in our understanding of RSTS during the last decade, although, as highlighted in the previous section, important questions remain. The best strategy to eradicate RSTS and other congenital IDDs arguably passes for the identification of the whole panel of mutations responsible of the disease and the refinement and widespread use of early prenatal tests. However, in the meantime, the cure or treatment of these disorders will require a better understanding of the etiology of the disease, which is particularly challenging in multifunction proteins like CBP or p300 with a large number of downstream substrates. Future studies in animal models should further dissect the developmental or adult origin of cognitive impairments, being particularly relevant the efforts aimed to rescue specific defects in adults through genetic or pharmacological approaches. The positive results in this arena would encourage the translation of these strategies to humans bringing hope to the treatment of ID in this today incurable disease.

As described in the previous sections, other IDDs are caused by histone-modifying enzymes that interplay with CBP/p300 activity and affect common gene networks. Therefore, the progress in the understanding of the molecular etiology of RSTS and the role of the proteins CBP and p300 in brain development and function goes well beyond RSTS and can also contribute to the understanding and therapy of other IDDs and answer fundamental questions in cell biology.

ACKNOWLEDGMENTS

Authors' research is supported by Grants SAF2011-22855 and SAF2011-22506 from the Spanish Ministry of Science and Innovation and Prometeo/2012/005 from the Generalitat Valenciana. The Instituto de Neurociencias is a "Centre of Excellence Severo Ochoa".

REFERENCES

1. Kleefstra T, Schenck A, Kramer JM, van Bokhoven H. The genetics of cognitive epigenetics. *Neuropharmacology*. 2014;80:83–94.
2. Bayes A, van de Lagemaat LN, Collins MO, et al. Characterization of the proteome, diseases and evolution of the human postsynaptic density. *Nat Neurosci*. 2011;14:19–21.
3. Padfield CJ, Partington MW, Simpson NE. The Rubinstein–Taybi syndrome. *Arch Dis Child*. 1968;43:94–101.
4. Berdasco M, Esteller M. Genetic syndromes caused by mutations in epigenetic genes. *Hum Genet*. 2013;132:359–383.
5. Goldberg AD, Allis CD, Bernstein E. Epigenetics: a landscape takes shape. *Cell*. 2007;128:635–638.
6. Hsieh J, Gage FH. Epigenetic control of neural stem cell fate. *Curr Opin Genet Dev*. 2004;14:461–469.
7. Lunyak VV, Rosenfeld MG. Epigenetic regulation of stem cell fate. *Hum Mol Genet*. 2008;17:R28–R36.
8. Gifford CA, Ziller MJ, Gu H, et al. Transcriptional and epigenetic dynamics during specification of human embryonic stem cells. *Cell*. 2013;153:1149–1163.

9. Xie W, Schultz MD, Lister R, et al. Epigenomic analysis of multilineage differentiation of human embryonic stem cells. *Cell*. 2013;153:1134–1148.
10. Zhu J, Adli M, Zou JY, et al. Genome-wide chromatin state transitions associated with developmental and environmental cues. *Cell*. 2013;152:642–654.
11. Liang G, Zhang Y. Genetic and epigenetic variations in iPSCs: potential causes and implications for application. *Cell Stem Cell*. 2013;13:149–159.
12. Makrythanasis P, van Bon BW, Steehouwer M, et al. MLL2 mutation detection in 86 patients with kabuki syndrome: a genotype–phenotype study. *Clin Genet*. 2013;84(6):539–545.
13. Miyake N, Mizuno S, Okamoto N, et al. KDM6A point mutations cause Kabuki syndrome. *Hum Mutat*. 2013;34:108–110.
14. Parkel S, Lopez-Atalaya JP, Barco A. Histone H3 lysine methylation in cognition and intellectual disability disorders. *Learn Mem*. 2013;20:570–579.
15. Kleefstra T, Kramer JM, Neveling K, et al. Disruption of an EHMT1-associated chromatin-modification module causes intellectual disability. *Am J Hum Genet*. 2012;91:73–82.
16. Rubinstein JH, Taybi H. Broad thumbs and toes and facial abnormalities. A possible mental retardation syndrome. *Am J Dis Child*. 1963;105:588–608.
17. Hennekam RC. Rubinstein–Taybi syndrome. *Eur J Hum Genet*. 2006;14:981–985.
18. Park E, Kim Y, Ryu H, Kowall NW, Lee J, Ryu H. Epigenetic mechanisms of Rubinstein–Taybi syndrome. *Neuromolecul Med*. 2014;16:16–24.
19. Roelfsema JH, Peters DJ. Rubinstein–Taybi syndrome: clinical and molecular overview. *Expert Rev Mol Med*. 2007;9:1–16.
20. Wiley S, Swayne S, Rubinstein JH, Lanphear NE, Stevens CA. Rubinstein–Taybi syndrome medical guidelines. *Am J Med Genet A*. 2003;119:101–110.
21. Hennekam RC, Baselier AC, Beyaert E, et al. Psychological and speech studies in Rubinstein–Taybi syndrome. *Am J Ment Retard*. 1992;96:645–660.
22. Schorry EK, Keddache M, Lanphear N, et al. Genotype–phenotype correlations in Rubinstein–Taybi syndrome. *Am J Med Genet A*. 2008;146A:2512–2519.
23. Bourdeaut F, Miquel C, Richer W, et al. Rubinstein–Taybi syndrome predisposing to non-WNT, non-SHH, group 3 medulloblastoma. *Pediatr Blood Cancer*. 2014;61:383–386.
24. Bartsch O, Kress W, Kempf O, Lechno S, Haaf T, Zechner U. Inheritance and variable expression in Rubinstein–Taybi syndrome. *Am J Med Genet A*. 2010;152A:2254–2261.
25. Petrij F, Giles RH, Dauwerse HG, et al. Rubinstein–Taybi syndrome caused by mutations in the transcriptional co-activator CBP. *Nature*. 1995;376:348–351.
26. Chrivia JC, Kwok RP, Lamb N, Hagiwara M, Montminy MR, Goodman RH. Phosphorylated CREB binds specifically to the nuclear protein CBP. *Nature*. 1993;365:855–859.
27. Roelfsema JH, White SJ, Ariyurek Y, et al. Genetic heterogeneity in Rubinstein–Taybi syndrome: mutations in both the CBP and EP300 genes cause disease. *Am J Hum Genet*. 2005;76:572–580.
28. Egan C, Jelsma TN, Howe JA, Bayley ST, Ferguson B, Branton PE. Mapping of cellular protein-binding sites on the products of early-region 1A of human adenovirus type 5. *Mol Cell Biol*. 1988;8:3955–3959.
29. Whyte P, Williamson NM, Harlow E. Cellular targets for transformation by the adenovirus E1A proteins. *Cell*. 1989;56:67–75.
30. Negri G, Milani D, Colapietro P, et al. Clinical and molecular characterization of Rubinstein–Taybi syndrome patients carrying distinct novel mutations of the EP300 gene. *Clin Genet*. 2014. http://dx.doi.org/10.1111/cge.12348. [Epub ahead of print].
31. Bartholdi D, Roelfsema JH, Papadia F, et al. Genetic heterogeneity in Rubinstein–Taybi syndrome: delineation of the phenotype of the first patients carrying mutations in EP300. *J Med Genet*. 2007;44:327–333.

32. Bartsch O, Labonte J, Albrecht B, et al. Two patients with EP300 mutations and facial dysmorphism different from the classic Rubinstein–Taybi syndrome. *Am J Med Genet A*. 2010;152A:181–184.
33. Foley P, Bunyan D, Stratton J, Dillon M, Lynch SA. Further case of Rubinstein–Taybi syndrome due to a deletion in EP300. *Am J Med Genet A*. 2009;149A(5):997–1000.
34. Zimmermann N, Acosta AM, Kohlhase J, Bartsch O. Confirmation of EP300 gene mutations as a rare cause of Rubinstein–Taybi syndrome. *Eur J Hum Genet*. 2007;15:837–842.
35. Coupry I, Roudaut C, Stef M, et al. Molecular analysis of the CBP gene in 60 patients with Rubinstein–Taybi syndrome. *J Med Genet*. 2002;39:415–421.
36. Bartsch O, Wagner A, Hinkel GK, et al. FISH studies in 45 patients with Rubinstein–Taybi syndrome: deletions associated with polysplenia, hypoplastic left heart and death in infancy. *Eur J Hum Genet*. 1999;7:748–756.
37. Bartsch O, Rasi S, Delicado A, et al. Evidence for a new contiguous gene syndrome, the chromosome 16p13.3 deletion syndrome alias severe Rubinstein–Taybi syndrome. *Hum Genet*. 2006;120:179–186.
38. Bartsch O, Locher K, Meinecke P, et al. Molecular studies in 10 cases of Rubinstein–Taybi syndrome, including a mild variant showing a missense mutation in codon 1175 of CREBBP. *J Med Genet*. 2002;39:496–501.
39. Thienpont B, Bena F, Breckpot J, et al. Duplications of the critical Rubinstein–Taybi deletion region on chromosome 16p13.3 cause a novel recognisable syndrome. *J Med Genet*. 2010;47:155–161.
40. Iyer NG, Ozdag H, Caldas C. p300/CBP and cancer. *Oncogene*. 2004;23:4225–4231.
41. Mullighan CG, Zhang J, Kasper LH, et al. CREBBP mutations in relapsed acute lymphoblastic leukaemia. *Nature*. 2011;471:235–239.
42. Pasqualucci L, Dominguez-Sola D, Chiarenza A, et al. Inactivating mutations of acetyltransferase genes in B-cell lymphoma. *Nature*. 2011;471:189–195.
43. Peifer M, Fernandez-Cuesta L, Sos ML, et al. Integrative genome analyses identify key somatic driver mutations of small-cell lung cancer. *Nat Genet*. 2012;44:1104–1110.
44. Hood RL, Lines MA, Nikkel SM, et al. Mutations in SRCAP, encoding SNF2-related CREBBP activator protein, cause Floating-Harbor syndrome. *Am J Hum Genet*. 2012;90:308–313.
45. Lacombe D, Patton MA, Elleau C, Battin J. Floating-Harbor syndrome: description of a further patient, review of the literature, and suggestion of autosomal dominant inheritance. *Eur J Pediatr*. 1995;154:658–661.
46. Le Goff C, Mahaut C, Bottani A, et al. Not all Floating–Harbor syndrome cases are due to mutations in exon 34 of SRCAP. *Hum Mutat*. 2013;34:88–92.
47. Guo W, Crossey EL, Zhang L, et al. Alcohol exposure decreases CREB binding protein expression and histone acetylation in the developing cerebellum. *PLoS One*. 2011;6:e19351.
48. Bedford DC, Kasper LH, Fukuyama T, Brindle PK. Target gene context influences the transcriptional requirement for the KAT3 family of CBP and p300 histone acetyltransferases. *Epigenetics*. 2010;5:9–15.
49. Kasper LH, Fukuyama T, Biesen MA, et al. Conditional knockout mice reveal distinct functions for the global transcriptional coactivators CBP and p300 in T-cell development. *Mol Cell Biol*. 2006;26:789–809.
50. Heintzman ND, Hon GC, Hawkins RD, et al. Histone modifications at human enhancers reflect global cell-type-specific gene expression. *Nature*. 2009;459: 108–112.
51. Kim TK, Hemberg M, Gray JM, et al. Widespread transcription at neuronal activity-regulated enhancers. *Nature*. 2010;465:182–187.
52. Visel A, Blow MJ, Li Z, et al. ChIP-seq accurately predicts tissue-specific activity of enhancers. *Nature*. 2009;457:854–858.

53. Wang Z, Zang C, Cui K, et al. Genome-wide mapping of HATs and HDACs reveals distinct functions in active and inactive genes. *Cell.* 2009;138:1019–1031.
54. Valor LM, Viosca J, Lopez-Atalaya JP, Barco A. Lysine acetyltransferases CBP and p300 as therapeutic targets in cognitive and neurodegenerative disorders. *Curr Pharm Des.* 2013;19:5051–5064.
55. Janknecht R. The versatile functions of the transcriptional coactivators p300 and CBP and their roles in disease. *Histol Histopathol.* 2002;17:657–668.
56. Kalkhoven E. CBP and p300: HATs for different occasions. *Biochem Pharmacol.* 2004;68:1145–1155.
57. Wang L, Tang Y, Cole PA, Marmorstein R. Structure and chemistry of the p300/CBP and Rtt109 histone acetyltransferases: implications for histone acetyltransferase evolution and function. *Curr Opin Struct Biol.* 2008;18:741–747.
58. Choudhary C, Kumar C, Gnad F, et al. Lysine acetylation targets protein complexes and co-regulates major cellular functions. *Science.* 2009;325:834–840.
59. Henriksen P, Wagner SA, Weinert BT, et al. Proteome-wide analysis of lysine acetylation suggests its broad regulatory scope in *Saccharomyces cerevisiae. Mol Cell Proteomics.* 2012;11:1510–1522.
60. Kim SC, Sprung R, Chen Y, et al. Substrate and functional diversity of lysine acetylation revealed by a proteomics survey. *Mol Cell.* 2006;23:607–618.
61. Lundby A, Lage K, Weinert BT, et al. Proteomic analysis of lysine acetylation sites in rat tissues reveals organ specificity and subcellular patterns. *Cell Rep.* 2012;2:419–431.
62. Zhao S, Xu W, Jiang W, et al. Regulation of cellular metabolism by protein lysine acetylation. *Science.* 2010;327:1000–1004.
63. Allis CD, Berger SL, Cote J, et al. New nomenclature for chromatin-modifying enzymes. *Cell.* 2007;131:633–636.
64. Kouzarides T. Chromatin modifications and their function. *Cell.* 2007;128:693–705.
65. Turner BM. The adjustable nucleosome: an epigenetic signaling module. *Trends Genet.* 2012;28:436–444.
66. Zhang H, Gao L, Anandhakumar J, Gross DS. Uncoupling transcription from covalent histone modification. *PLoS Genet.* 2014;10:e1004202.
67. Chen L, Fischle W, Verdin E, Greene WC. Duration of nuclear NF-kappaB action regulated by reversible acetylation. *Science.* 2001;293:1653–1657.
68. Tang Y, Zhao W, Chen Y, Zhao Y, Gu W. Acetylation is indispensable for p53 activation. *Cell.* 2008;133:612–626.
69. Schroder S, Herker E, Itzen F, et al. Acetylation of RNA polymerase II regulates growth-factor-induced gene transcription in mammalian cells. *Mol Cell.* 2013;52:314–324.
70. Chan HM, La Thangue NB. p300/CBP proteins: HATs for transcriptional bridges and scaffolds. *J Cell Sci.* 2001;114:2363–2373.
71. Tanaka Y, Naruse I, Maekawa T, Masuya H, Shiroishi T, Ishii S. Abnormal skeletal patterning in embryos lacking a single Cbp allele: a partial similarity with Rubinstein–Taybi syndrome. *Proc Natl Acad Sci USA.* 1997;94:10215–10220.
72. Kung AL, Rebel VI, Bronson RT, et al. Gene dose-dependent control of hematopoiesis and hematologic tumor suppression by CBP. *Genes Dev.* 2000;14:272–277.
73. Yao TP, Oh SP, Fuchs M, et al. Gene dosage-dependent embryonic development and proliferation defects in mice lacking the transcriptional integrator p300. *Cell.* 1998;93:361–372.
74. Oike Y, Hata A, Mamiya T, et al. Truncated CBP protein leads to classical Rubinstein–Taybi syndrome phenotypes in mice: implications for a dominant-negative mechanism. *Hum Mol Genet.* 1999;8:387–396.
75. Roth JF, Shikama N, Henzen C, et al. Differential role of p300 and CBP acetyltransferase during myogenesis: p300 acts upstream of MyoD and Myf5. *EMBO J.* 2003;22:5186–5196.

76. Kasper LH, Boussouar F, Ney PA, et al. A transcription-factor-binding surface of coactivator p300 is required for haematopoiesis. *Nature*. 2002;419:738–743.
77. Kasper LH, Boussouar F, Boyd K, et al. Two transactivation mechanisms cooperate for the bulk of HIF-1-responsive gene expression. *EMBO J*. 2005;24:3846–3858.
78. Zhang Z, Hofmann C, Casanova E, Schutz G, Lutz B. Generation of a conditional allele of the CBP gene in mouse. *Genesis*. 2004;40:82–89.
79. Kang-Decker N, Tong C, Boussouar F, et al. Loss of CBP causes T cell lymphomagenesis in synergy with p27Kip1 insufficiency. *Cancer Cell*. 2004;5:177–189.
80. Korzus E, Rosenfeld MG, Mayford M. CBP histone acetyltransferase activity is a critical component of memory consolidation. *Neuron*. 2004;42:961–972.
81. Wood MA, Kaplan MP, Park A, et al. Transgenic mice expressing a truncated form of CREB-binding protein (CBP) exhibit deficits in hippocampal synaptic plasticity and memory storage. *Learn Mem*. 2005;12:111–119.
82. Oliveira AM, Wood MA, McDonough CB, Abel T. Transgenic mice expressing an inhibitory truncated form of p300 exhibit long-term memory deficits. *Learn Mem*. 2007;14:564–572.
83. Shikama N, Lutz W, Kretzschmar R, et al. Essential function of p300 acetyltransferase activity in heart, lung and small intestine formation. *EMBO J*. 2003;22:5175–5185.
84. Bedford DC, Kasper LH, Wang R, Chang Y, Green DR, Brindle PK. Disrupting the CH1 domain structure in the acetyltransferases CBP and p300 results in lean mice with increased metabolic control. *Cell Metab*. 2011;14:219–230.
85. Erdmann G, Schutz G, Berger S. Inducible gene inactivation in neurons of the adult mouse forebrain. *BMC Neurosci*. 2007;8:63.
86. Gomez RA, Pentz ES, Jin X, Cordaillat M, Sequeira Lopez ML. CBP and p300 are essential for renin cell identity and morphological integrity of the kidney. *Am J Physiol Heart Circ Physiol*. 2009;296:H1255–H1262.
87. Chen G, Zou X, Watanabe H, van Deursen JM, Shen J. CREB binding protein is required for both short-term and long-term memory formation. *J Neurosci*. 2010;30:13066–13077.
88. Valor LM, Pulopulos MM, Jimenez-Minchan M, Olivares R, Lutz B, Barco A. Ablation of CBP in forebrain principal neurons causes modest memory and transcriptional defects and a dramatic reduction of histone acetylation, but does not affect cell viability. *J Neurosci*. 2011;31:1652–1663.
89. Barrett RM, Malvaez M, Kramar E, et al. Hippocampal focal knockout of CBP affects specific histone modifications, long-term potentiation, and long-term memory. *Neuropsychopharmacology*. 2011;36(8):1545–1556.
90. Malvaez M, Mhillaj E, Matheos DP, Palmery M, Wood MA. CBP in the nucleus accumbens regulates cocaine-induced histone acetylation and is critical for cocaine-associated behaviors. *J Neurosci*. 2011;31:16941–16948.
91. Mayford M, Bach ME, Huang YY, Wang L, Hawkins RD, Kandel ER. Control of memory formation through regulated expression of a CaMKII transgene. *Science*. 1996;274:1678–1683.
92. Goodman RH, Smolik S. CBP/p300 in cell growth, transformation, and development. *Genes Dev*. 2000;14:1553–1577.
93. Tanaka Y, Naruse I, Hongo T, et al. Extensive brain hemorrhage and embryonic lethality in a mouse null mutant of CREB-binding protein. *Mech Dev*. 2000;95:133–145.
94. Viosca J, Lopez-Atalaya JP, Olivares R, Eckner R, Barco A. Syndromic features and mild cognitive impairment in mice with genetic reduction on p300 activity: differential contribution of p300 and CBP to Rubinstein–Taybi syndrome etiology. *Neurobiol Dis*. 2010;37:186–194.
95. Phan HM, Xu AW, Coco C, et al. GCN5 and p300 share essential functions during early embryogenesis. *Dev Dyn*. 2005;233:1337–1347.

96. Parker D, Ferreri K, Nakajima T, et al. Phosphorylation of CREB at Ser-133 induces complex formation with CREB-binding protein via a direct mechanism. *Mol Cell Biol.* 1996;16:694–703.
97. Bhattacherjee V, Horn KH, Singh S, Webb CL, Pisano MM, Greene RM. CBP/p300 and associated transcriptional co-activators exhibit distinct expression patterns during murine craniofacial and neural tube development. *Int J Dev Biol.* 2009;53:1097–1104.
98. Partanen A, Motoyama J, Hui CC. Developmentally regulated expression of the transcriptional cofactors/histone acetyltransferases CBP and p300 during mouse embryogenesis. *Int J Dev Biol.* 1999;43:487–494.
99. Wang J, Weaver IC, Gauthier-Fisher A, et al. CBP histone acetyltransferase activity regulates embryonic neural differentiation in the normal and Rubinstein–Taybi syndrome brain. *Dev Cell.* 2010;18:114–125.
100. Tsui D, Voronova A, Gallagher D, Kaplan DR, Miller FD, Wang J. CBP regulates the differentiation of interneurons from ventral forebrain neural precursors during murine development. *Dev Biol.* 2014;385:230–241.
101. Cantani A, Gagliesi D. Rubinstein–Taybi syndrome. Review of 732 cases and analysis of the typical traits. *Eur Rev Med Pharmacol Sci.* 1998;2:81–87.
102. Akimaru H, Chen Y, Dai P, et al. Drosophila CBP is a co-activator of cubitus interruptus in hedgehog signalling. *Nature.* 1997;386:735–738.
103. Holmqvist PH, Boija A, Philip P, Crona F, Stenberg P, Mannervik M. Preferential genome targeting of the CBP co-activator by Rel and smad proteins in early *Drosophila melanogaster* embryos. *PLoS Genet.* 2012;8:e1002769.
104. Kumar JP, Jamal T, Doetsch A, Turner FR, Duffy JB. CREB binding protein functions during successive stages of eye development in Drosophila. *Genetics.* 2004;168: 877–893.
105. Schmidt I, Franzdottir SR, Edenfeld G, Rodrigues F, Zierau A, Klambt C. Transcriptional regulation of peripheral glial cell differentiation in the embryonic nervous system of Drosophila. *Glia.* 2011;59:1264–1272.
106. Shim JH, Greenblatt MB, Singh A, et al. Administration of BMP2/7 in utero partially reverses Rubinstein–Taybi syndrome-like skeletal defects induced by Pdk1 or Cbp mutations in mice. *J Clin Invest.* 2012;122:91–106.
107. Yamauchi T, Oike Y, Kamon J, et al. Increased insulin sensitivity despite lipodystrophy in crebbp heterozygous mice. *Nat Genet.* 2002;30:221–226.
108. Wang F, Marshall CB, Ikura M. Transcriptional/epigenetic regulator CBP/p300 in tumorigenesis: structural and functional versatility in target recognition. *Cell Mol Life Sci.* 2013;70:3989–4008.
109. Miller RW, Rubinstein JH. Tumors in Rubinstein–Taybi syndrome. *Am J Med Genet.* 1995;56:112–115.
110. Barco A. The Rubinstein–Taybi syndrome: modeling mental impairment in the mouse. *Genes Brain Behav.* 2007;6(Suppl 1):32–39.
111. Levine AA, Guan Z, Barco A, Xu S, Kandel ER, Schwartz JH. CREB-binding protein controls response to cocaine by acetylating histones at the fosB promoter in the mouse striatum. *Proc Natl Acad Sci USA.* 2005;102:19186–19191.
112. Alarcon JM, Malleret G, Touzani K, et al. Chromatin acetylation, memory, and LTP are impaired in CBP+/− mice: a model for the cognitive deficit in Rubinstein–Taybi syndrome and its amelioration. *Neuron.* 2004;42:947–959.
113. Oliveira AM, Abel T, Brindle PK, Wood MA. Differential role for CBP and p300 CREB-binding domain in motor skill learning. *Behav Neurosci.* 2006;120:724–729.
114. Gotts EE, Liemohn WP. Behavioral characteristics of three children with the broad thumb–hallux (Rubinstein–Taybi) syndrome. *Biol Psychiatry.* 1977;12:413–423.
115. Oliveira AM, Estevez MA, Hawk JD, Grimes S, Brindle PK, Abel T. Subregion-specific p300 conditional knock-out mice exhibit long-term memory impairments. *Learn Mem.* 2011;18:161–169.

116. Lopez-Atalaya JP, Ciccarelli A, Viosca J, et al. CBP is required for environmental enrichment-induced neurogenesis and cognitive enhancement. *EMBO J.* 2011;30: 4287–4298.
117. Drew LJ, Fusi S, Hen R. Adult neurogenesis in the mammalian hippocampus: why the dentate gyrus? *Learn Mem.* 2013;20:710–729.
118. Ming GL, Song H. Adult neurogenesis in the mammalian brain: significant answers and significant questions. *Neuron.* 2011;70:687–702.
119. Zhao C, Deng W, Gage FH. Mechanisms and functional implications of adult neurogenesis. *Cell.* 2008;132:645–660.
120. Aimone JB, Deng W, Gage FH. Resolving new memories: a critical look at the dentate gyrus, adult neurogenesis, and pattern separation. *Neuron.* 2011;70:589–596.
121. Clelland CD, Choi M, Romberg C, et al. A functional role for adult hippocampal neurogenesis in spatial pattern separation. *Science.* 2009;325:210–213.
122. Sahay A, Scobie KN, Hill AS, et al. Increasing adult hippocampal neurogenesis is sufficient to improve pattern separation. *Nature.* 2011;472:466–470.
123. Deng W, Aimone JB, Gage FH. New neurons and new memories: how does adult hippocampal neurogenesis affect learning and memory? *Nat Rev Neurosci.* 2010;11: 339–350.
124. Sahay A, Wilson DA, Hen R. Pattern separation: a common function for new neurons in hippocampus and olfactory bulb. *Neuron.* 2011;70:582–588.
125. Ma DK, Kim WR, Ming GL, Song H. Activity-dependent extrinsic regulation of adult olfactory bulb and hippocampal neurogenesis. *Ann N Y Acad Sci.* 2009;1170:664–673.
126. Ma DK, Marchetto MC, Guo JU, Ming GL, Gage FH, Song H. Epigenetic choreographers of neurogenesis in the adult mammalian brain. *Nat Neurosci.* 2010;13:1338–1344.
127. Zhang RR, Cui QY, Murai K, et al. Tet1 regulates adult hippocampal neurogenesis and cognition. *Cell Stem Cell.* 2013;13:237–245.
128. Wang J, Gallagher D, DeVito LM, et al. Metformin activates an atypical PKC-CBP pathway to promote neurogenesis and enhance spatial memory formation. *Cell Stem Cell.* 2012;11:23–35.
129. Denis-Donini S, Dellarole A, Crociara P, et al. Impaired adult neurogenesis associated with short-term memory defects in NF-kappaB p50-deficient mice. *J Neurosci.* 2008;28:3911–3919.
130. Medrano S, Scrable H. Maintaining appearances—the role of p53 in adult neurogenesis. *Biochem Biophys Res Commun.* 2005;331:828–833.
131. Lee S, Lee SK. Crucial roles of histone-modifying enzymes in mediating neural cell-type specification. *Curr Opin Neurobiol.* 2010;20:29–36.
132. Chatterjee S, Mizar P, Cassel R, et al. A novel activator of CBP/p300 acetyltransferases promotes neurogenesis and extends memory duration in adult mice. *J Neurosci.* 2013;33:10698–10712.
133. Hao Y, Creson T, Zhang L, et al. Mood stabilizer valproate promotes ERK pathway-dependent cortical neuronal growth and neurogenesis. *J Neurosci.* 2004;24:6590–6599.
134. Hsieh J, Nakashima K, Kuwabara T, Mejia E, Gage FH. Histone deacetylase inhibition-mediated neuronal differentiation of multipotent adult neural progenitor cells. *Proc Natl Acad Sci USA.* 2004;101:16659–16664.
135. Umka J, Mustafa S, ElBeltagy M, et al. Valproic acid reduces spatial working memory and cell proliferation in the hippocampus. *Neuroscience.* 2010;166:15–22.
136. Yu IT, Park JY, Kim SH, Lee JS, Kim YS, Son H. Valproic acid promotes neuronal differentiation by induction of proneural factors in association with H4 acetylation. *Neuropharmacology.* 2009;56:473–480.
137. Barco A, Kandel ER. The role of CREB and CBP in brain function. In: Thiel G, ed. *Transcription factors in the nervous system: development, brain functions, and diseases.* Weinheim, FRG: Wiley-VCH Verlag GmbH & Co. KGaA; 2006.

138. Barco A, Marie H. Genetic approaches to investigate the role of CREB in neuronal plasticity and memory. *Mol Neurobiol.* 2011;44:330–349.
139. Benito E, Barco A. CREB's control of intrinsic and synaptic plasticity: implications for CREB-dependent memory models. *Trends Neurosci.* 2010;33:230–240.
140. Kalkhoven E, Roelfsema JH, Teunissen H, et al. Loss of CBP acetyltransferase activity by PHD finger mutations in Rubinstein–Taybi syndrome. *Hum Mol Genet.* 2003;12:441–450.
141. Wood MA, Attner MA, Oliveira AM, Brindle PK, Abel T. A transcription factor-binding domain of the coactivator CBP is essential for long-term memory and the expression of specific target genes. *Learn Mem.* 2006;13:609–617.
142. Lundblad JR, Kwok RP, Laurance ME, Harter ML, Goodman RH. Adenoviral E1A-associated protein p300 as a functional homologue of the transcriptional co-activator CBP. *Nature.* 1995;374:85–88.
143. Graff J, Tsai LH. Histone acetylation: molecular mnemonics on the chromatin. *Nat Rev Neurosci.* 2013;14:97–111.
144. Peixoto L, Abel T. The role of histone acetylation in memory formation and cognitive impairments. *Neuropsychopharmacology.* 2013;38:62–76.
145. Lopez-Atalaya JP, Gervasini C, Mottadelli F, et al. Histone acetylation deficits in lymphoblastoid cell lines from patients with Rubinstein–Taybi syndrome. *J Med Genet.* 2012;49:66–74.
146. Lee J, Hagerty S, Cormier KA, et al. Monoallele deletion of CBP leads to pericentromeric heterochromatin condensation through ESET expression and histone H3 (K9) methylation. *Hum Mol Genet.* 2008;17:1774–1782.
147. Fischer A, Sananbenesi F, Mungenast A, Tsai LH. Targeting the correct HDAC(s) to treat cognitive disorders. *Trends Pharmacol Sci.* 2010;31:605–617.
148. Graff J, Tsai LH. The potential of HDAC inhibitors as cognitive enhancers. *Annu Rev Pharmacol Toxicol.* 2013;53:311–330.
149. Bolden JE, Peart MJ, Johnstone RW. Anticancer activities of histone deacetylase inhibitors. *Nat Rev Drug Discov.* 2006;5:769–784.
150. Bredy TW, Wu H, Crego C, Zellhoefer J, Sun YE, Barad M. Histone modifications around individual BDNF gene promoters in prefrontal cortex are associated with extinction of conditioned fear. *Learn Mem.* 2007;14:268–276.
151. Guan JS, Haggarty SJ, Giacometti E, et al. HDAC2 negatively regulates memory formation and synaptic plasticity. *Nature.* 2009;459:55–60.
152. Levenson JM, O'Riordan KJ, Brown KD, Trinh MA, Molfese DL, Sweatt JD. Regulation of histone acetylation during memory formation in the hippocampus. *J Biol Chem.* 2004;279:40545–40559.
153. McQuown SC, Barrett RM, Matheos DP, et al. HDAC3 is a critical negative regulator of long-term memory formation. *J Neurosci.* 2011;31:764–774.
154. Roozendaal B, Hernandez A, Cabrera SM, et al. Membrane-associated glucocorticoid activity is necessary for modulation of long-term memory via chromatin modification. *J Neurosci.* 2010;30:5037–5046.
155. Stefanko DP, Barrett RM, Ly AR, Reolon GK, Wood MA. Modulation of long-term memory for object recognition via HDAC inhibition. *Proc Natl Acad Sci USA.* 2009;106:9447–9452.
156. Vecsey CG, Hawk JD, Lattal KM, et al. Histone deacetylase inhibitors enhance memory and synaptic plasticity via CREB:CBP-dependent transcriptional activation. *J Neurosci.* 2007;27:6128–6140.
157. Yeh SH, Lin CH, Gean PW. Acetylation of nuclear factor-kappaB in rat amygdala improves long-term but not short-term retention of fear memory. *Mol Pharmacol.* 2004;65:1286–1292.

158. Castellano JF, Fletcher BR, Patzke H, et al. Reassessing the effects of histone deacetylase inhibitors on hippocampal memory and cognitive aging. *Hippocampus*. 2014;24(8):1006–1016.
159. Graff J, Joseph NF, Horn ME, et al. Epigenetic priming of memory updating during reconsolidation to attenuate remote fear memories. *Cell*. 2014;156:261–276.
160. Fischer A, Sananbenesi F, Wang X, Dobbin M, Tsai LH. Recovery of learning and memory is associated with chromatin remodelling. *Nature*. 2007;447:178–182.
161. Wood MA, Hawk JD, Abel T. Combinatorial chromatin modifications and memory storage: a code for memory? *Learn Mem*. 2006;13:241–244.
162. Haettig J, Stefanko DP, Multani ML, Figueroa DX, McQuown SC, Wood MA. HDAC inhibition modulates hippocampus-dependent long-term memory for object location in a CBP-dependent manner. *Learn Mem*. 2011;18:71–79.
163. West AC, Johnstone RW. New and emerging HDAC inhibitors for cancer treatment. *J Clin Invest*. 2014;124:30–39.
164. Lopez-Atalaya JP, Ito S, Valor LM, Benito E, Barco A. Genomic targets, and histone acetylation and gene expression profiling of neural HDAC inhibition. *Nucleic Acids Res*. 2013;41:8072–8084.
165. Boswell-Smith V, Spina D, Page CP. Phosphodiesterase inhibitors. *Br J Pharmacol*. 2006;147(Suppl 1):S252–S257.
166. Scott AI, Perini AF, Shering PA, Whalley LJ. In-patient major depression: is rolipram as effective as amitriptyline? *Eur J Clin Pharmacol*. 1991;40:127–129.
167. Barad M, Bourtchouladze R, Winder DG, Golan H, Kandel E. Rolipram, a type IV-specific phosphodiesterase inhibitor, facilitates the establishment of long-lasting long-term potentiation and improves memory. *Proc Natl Acad Sci USA*. 1998;95: 15020–15025.
168. Zhang HT, Crissman AM, Dorairaj NR, Chandler LJ, O'Donnell JM. Inhibition of cyclic AMP phosphodiesterase (PDE4) reverses memory deficits associated with NMDA receptor antagonism. *Neuropsychopharmacology*. 2000;23:198–204.
169. Zhang HT, Huang Y, Suvarna NU, et al. Effects of the novel PDE4 inhibitors MEM1018 and MEM1091 on memory in the radial-arm maze and inhibitory avoidance tests in rats. *Psychopharmacology (Berl)*. 2005;179:613–619.
170. Li YF, Huang Y, Amsdell SL, Xiao L, O'Donnell JM, Zhang HT. Antidepressant- and anxiolytic-like effects of the phosphodiesterase-4 inhibitor rolipram on behavior depend on cyclic AMP response element binding protein-mediated neurogenesis in the hippocampus. *Neuropsychopharmacology*. 2009;34:2404–2419.
171. Bourtchouladze R, Lidge R, Catapano R, et al. A mouse model of Rubinstein–Taybi syndrome: defective long-term memory is ameliorated by inhibitors of phosphodiesterase 4. *Proc Natl Acad Sci USA*. 2003;100:10518–10522.
172. Nithianantharajah J, Hannan AJ. Enriched environments, experience-dependent plasticity and disorders of the nervous system. *Nat Rev Neurosci*. 2006;7:697–709.
173. Duman RS, Nakagawa S, Malberg J. Regulation of adult neurogenesis by antidepressant treatment. *Neuropsychopharmacology*. 2001;25:836–844.
174. van Praag H, Kempermann G, Gage FH. Running increases cell proliferation and neurogenesis in the adult mouse dentate gyrus. *Nat Neurosci*. 1999;2:266–270.
175. van Praag H, Shubert T, Zhao C, Gage FH. Exercise enhances learning and hippocampal neurogenesis in aged mice. *J Neurosci*. 2005;25:8680–8685.
176. Sweatt JD. The emerging field of neuroepigenetics. *Neuron*. 2013;80:624–632.
177. Borrelli E, Nestler EJ, Allis CD, Sassone-Corsi P. Decoding the epigenetic language of neuronal plasticity. *Neuron*. 2008;60:961–974.
178. Graff J, Kim D, Dobbin MM, Tsai LH. Epigenetic regulation of gene expression in physiological and pathological brain processes. *Physiol Rev*. 2011;91:603–649.

179. Guzman-Karlsson MC, Meadows JP, Gavin CF, Hablitz JJ, Sweatt JD. Transcriptional and epigenetic regulation of hebbian and non-hebbian plasticity. *Neuropharmacology*. 2014;80:3–17.
180. Riccio A. Dynamic epigenetic regulation in neurons: enzymes, stimuli and signaling pathways. *Nat Neurosci*. 2010;13:1330–1337.
181. Zocchi L, Sassone-Corsi P. Joining the dots: from chromatin remodeling to neuronal plasticity. *Curr Opin Neurobiol*. 2010;20:432–440.
182. Bedford DC, Brindle PK. Is histone acetylation the most important physiological function for CBP and p300? *Aging*. 2012;4:247–255.
183. Kasper LH, Lerach S, Wang J, Wu S, Jeevan T, Brindle PK. CBP/p300 double null cells reveal effect of coactivator level and diversity on CREB transactivation. *EMBO J*. 2010;29:3660–3672.
184. Obrietan K, Hoyt KR. CRE-mediated transcription is increased in Huntington's disease transgenic mice. *J Neurosci*. 2004;24:791–796.
185. Palop JJ, Chin J, Roberson ED, et al. Aberrant excitatory neuronal activity and compensatory remodeling of inhibitory hippocampal circuits in mouse models of Alzheimer's disease. *Neuron*. 2007;55:697–711.
186. Shimohata M, Shimohata T, Igarashi S, Naruse S, Tsuji S. Interference of CREB-dependent transcriptional activation by expanded polyglutamine stretches—augmentation of transcriptional activation as a potential therapeutic strategy for polyglutamine diseases. *J Neurochem*. 2005;93:654–663.
187. Shimohata T, Nakajima T, Yamada M, et al. Expanded polyglutamine stretches interact with TAFII130, interfering with CREB-dependent transcription. *Nat Genet*. 2000;26:29–36.
188. Shimohata T, Onodera O, Tsuji S. Interaction of expanded polyglutamine stretches with nuclear transcription factors leads to aberrant transcriptional regulation in polyglutamine diseases. *Neuropathology*. 2000;20:326–333.
189. Valor LM, Guiretti D. What's wrong with epigenetics in Huntington's disease? *Neuropharmacology*. 2014;80:103–114.
190. Gardner KE, Allis CD, Strahl BD. Operating on chromatin, a colorful language where context matters. *J Mol Biol*. 2011;409:36–46.
191. Henikoff S, Shilatifard A. Histone modification: cause or cog? *Trends Genet*. 2011;27:389–396.
192. Rando OJ. Combinatorial complexity in chromatin structure and function: revisiting the histone code. *Curr Opin Genet Dev*. 2012;22:148–155.
193. Xu W, Kasper LH, Lerach S, Jeevan T, Brindle PK. Individual CREB-target genes dictate usage of distinct cAMP-responsive coactivation mechanisms. *EMBO J*. 2007;26:2890–2903.
194. Jin Q, Yu LR, Wang L, et al. Distinct roles of GCN5/PCAF-mediated H3K9ac and CBP/p300-mediated H3K18/27ac in nuclear receptor transactivation. *EMBO J*. 2011;30:249–262.

CHAPTER SEVEN

The Intergenerational Effects of Early Adversity

Janina Galler, Danielle Galler Rabinowitz
Judge Baker Children's Center, Harvard Medical School, Boston, Massachusetts, USA

Contents

Abstract

Early insults during critical periods of brain development, both prenatal and postnatal, can result in epigenetic changes that may impact health and behavioral outcomes over the life span and into future generations. There is ample evidence that these early stages of brain development are sensitive to various environmental insults, including malnutrition, childhood trauma, and drug exposures. The notion that such changes, both physiological and behavioral, can also carry over into subsequent generations has long been recognized, especially in the context of experimental studies. However, epigenetic mechanisms capable of explaining such phenomena were not available until relatively recently, with most of this research published only within the last decade.

1. INTRODUCTION

This chapter sets out to identify the significant findings of intergenerational studies of the impact of various environmental factors on brain

Progress in Molecular Biology and Translational Science, Volume 128
ISSN 1877-1173
http://dx.doi.org/10.1016/B978-0-12-800977-2.00007-3

and behavior. First, we examine the available literature relating to research conducted on the effects on multiple generations of malnutrition, early adversity and trauma, and chemical exposures in both human populations and rat models. Second, we provide detailed results from the Barbados Nutrition Study and parallel rat studies in our laboratory. Third, we seek to highlight recent studies that have begun to look at underlying epigenetic mechanisms of such intergenerational transmission.

2. INTERGENERATIONAL STUDIES OF MALNUTRITION: HUMAN

Although the long-term effects of childhood malnutrition are now widely recognized,[1,2] studies documenting transgenerational effects of poor nutrition are less common. As early as 1949, Mussey considered: "It must be borne in mind that the diet of a given generation may effect several generations hence."[3] The recent advances in epigenetic research associated with adversity during early development have resulted in an increasing number of pertinent studies that suggest intergenerational effects of poor nutrition and other environmental factors. These include both studies in human populations and preclinical investigations in animal models.

The earliest studies of the transgenerational effects of poor nutrition were almost exclusively based on work with low-birth-weight infants within Western populations. More recently, this work has expanded to include studies in developing regions of the world. Although not always exclusively the result of poor nutrition, low birth weight is most prevalent in socioeconomically deprived populations and is thus closely associated with poverty. In the late 1950s, Sir Dugald Baird and his colleagues in Aberdeen, Scotland, followed a cohort of women who were short in stature and had an increased number of birth complications, likely associated with a history of inadequate nutrition during their own childhoods.[4–6] Follow-up studies of this cohort revealed that the offspring of these women were more likely to have lower IQ scores than children of mothers of average height.[7,8] One prevailing hypothesis was that maternal malnutrition resulted in aberrant pelvic shapes, ultimately translating to impaired reproductive competence. Although other possible mechanisms leading to transgenerational effects of poor nutrition were not recognized, this early study highlighted the important role of adequate maternal nutrition in preserving cognitive function from one generation to the next. Other studies, including those of Ounstead *et al.*,[9] Hackman *et al.*,[10] and Emanuel *et al.*[11] similarly reported that women who were of low birth weight were more likely to have offspring who were of low birth weight, thus affirming the risk of poor intergenerational outcomes with respect to infant size.[12]

The association between poor nutrition and birth weight has also been documented in epidemiological studies of survivors of the Dutch famine.[13] This series of studies showed that *in utero* exposure to malnutrition, especially during the second and third trimesters of pregnancy, was associated with decreased birth weights of infants even after correcting for maternal stature. Studies of transgenerational effects of birth weight have also been reported in developing countries, including Brazil, where maternal (but not paternal) birth weight and weight gain in early childhood was shown to be positively associated with birth weights in the next generation.[14] The persistence of low birth weight across generations was originally referred to as the intergenerational influences hypothesis.[15]

The Barker or the developmental origins of adult health and disease hypothesis,[12] as it is now referred to, states that low birth weight inevitably leads to a number of adult medical conditions including metabolic syndrome, cardiovascular disease,[16] and other health-related phenotypes, which are all programmed prenatally and modified by a limited nutrient supply during the early stages of development.[17] In their initial studies, Barker *et al.*[18,19] found that small, normal-term babies had an increased risk of hypertension and other cardiovascular diseases later in life, a finding that has been successfully replicated in a number of epidemiological studies.[20–22] The association between prenatal malnutrition, low birth weight, and adult-onset diabetes and obesity has also been reported among the survivors of the Dutch famine.[23] Although glucose intolerance was not increased in the adult survivors of the Siege of Leningrad, an increased prevalence of obesity and hypertension was identified.[24] In a Finnish cohort of adults born between 1933 and 1944, poor maternal nutrition and low birth weight were also correlated with an increased rate of obesity,[25] cardiovascular disease and associated mortality.[26,27] These medical conditions in adulthood may, in turn, compromise pregnancy and the *in utero* experience of the offspring, thereby contributing to the intergenerational transmission of malnutrition-related deficits. More recently, many of the low-birth-weight cohorts have been reassessed and have been found to have higher rates of mental health disorders and cognitive/behavioral impairments in adulthood.[28–30] These adverse mental health outcomes may directly impact the health and well-being of the next generation, as they contribute to impaired parenting and compromised educational, occupational, and socioeconomic opportunities.

In addition to studies of birth weight, the transgenerational effects of poor maternal nutrition have also been reported. A striking example of such effects was identified within a series of epidemiological studies by Kaati *et al.*, which examined nutrition-related factors among 15,000 Swedish men and women born between 1915 and 1929.[31] These investigators reported that the limited food supply available to the paternal grandparents resulted in their slowed growth during their childhood and periconception. This, in turn, significantly increased the risk of mortality (and diabetic deaths) of their grandchildren. These data also confirmed the sex-specific, male-line transmission of nutritional deficiency across generations. The 50-year follow-up of the grandchildren of the survivors of the Dutch famine revealed higher rates of neonatal adiposity[32] and higher BMIs,[33] both a result of maternal exposure to undernutrition. These findings corroborated earlier research findings by Lumey and Stein[34] who already identified that there may be long-term biological impacts into the next generation not directly corresponding to birth weight.

Complementing the studies on the implications of nutritional deficits, the effects of good nutrition during childhood have been shown to positively impact growth in the next generation. Protein supplementation of mothers and young children in a nutritionally and socioeconomically disadvantaged population in Guatemala, the INCAP study, improved the nutritional status of the supplemented women and children and also benefited the growth and development of the offspring of these children.[35,36] The protein-supplemented (atole) women gave birth to progeny, especially males, who were significantly taller than the children of mothers who received the no protein, low-energy (fresco) supplements. This study, therefore, clearly identified the potential reversibility of certain nutritional deficits over two generations.

3. LONG-TERM AND INTERGENERATIONAL EFFECTS OF POSTNATAL MALNUTRITION: THE BARBADOS NUTRITION STUDY

Although the vast majority of studies of the developmental origins of health and disease have focused on maternal malnutrition, there is ample evidence that poor nutrition in the postnatal period may also lead to long-term adverse effects on health, behavior, and cognition. Thus, studies that have followed stunted children or children with early childhood malnutrition have reported cognitive and behavioral deficits lasting to 22 years of age.[1,37,38] Similar mechanisms to those involved in perturbations associated with prenatal maternal undernutrition may be at play, including effects on programming, gene expression, and environmentally mediated epigenetic changes. Dobbing,[39] Morgane *et al.*,[40] and others[41] have long recognized the importance of the timing of insults during the prenatal and postnatal periods with respect to brain development in different species. Typically, adverse environmental exposures in the human, including nutritional deficits and other developmental insults, result in permanent changes to the brain if they occur during pregnancy through 2 years of age. Changes associated with insults occurring after this critical period of brain development are generally considered reversible.

Because of malnutrition's profound effects on neuronal growth, nutritional insults during infancy and early childhood put individuals at risk for lifelong impairment. These deficits can result in poor health, lower educational attainment, and diminished social status in adulthood. If the affected individuals themselves become parents, the impact of their early nutritional insult may even extend into the next generation. Thus, while there are numerous direct effects of childhood insult on brain and behavior, there may also be indirect effects of a social nature. Socioeconomic status is a powerful predictor of intellectual functioning and academic achievement, and the decreased social status of the parents has the potential to translate to intellectual compromise in their children, despite adequate nourishment throughout their development.

We consequently opted to examine the direct and indirect effects of childhood malnutrition in the context of the Barbados Nutrition Study, a 47-year longitudinal and now intergenerational study based on an epidemiological sample of children who had experienced an episode of moderate to severe protein–energy malnutrition in the first year of life. The development of these children was compared, using a case–control design, with that of healthy children recruited from the same neighborhoods and classrooms. Following their hospitalizations, study participants were enrolled in a nutritional intervention program that monitored their growth and development and assured adequate nutrition and good health from infancy to 12 years of age. Thus, for individuals initially identified as malnourished, our intervention guaranteed that their malnutrition episodes were limited to the first year of life; healthy controls had no growth failure or malnutrition throughout their lives. All participants were followed, at regular intervals, from early childhood to middle adulthood as part of their enrollment in the Barbados Nutrition Study.

Although the previously malnourished children achieved complete catch-up in their physical growth by adolescence,[42] we found that their cognitive and behavioral development remained adversely affected. A key finding from the Barbados Nutrition Study was the fourfold increase in attention deficits (from 15% to 60% of participants) throughout childhood, adolescence, and adulthood as compared with healthy controls.[43–45] In childhood and adolescence, the previously malnourished cohort also demonstrated lower IQ scores,[46] poor academic performance,[45] and increased mood and behavioral problems[47,48] relative to the control group. Differences in academic performance between groups were largely assessed through analysis of grade reports and scores on the national 11-plus examination, a test required for entrance into secondary schools on the island.[45]

When assessed at 40 years of age, the previously malnourished participants demonstrated continuing deficits in attention[49] and cognition.[50] In comparison with unaffected peers, the adults with histories of infant malnutrition had higher rates of attention deficits and lower IQ, approximately one standard deviation in magnitude, even after adjusting for the effects of childhood socioeconomic circumstances. In addition, approximately 25% of previously malnourished participants had IQs in the range of intellectual disability, and this prevalence rate was nine times higher than in the control group. Behavioral and neuropsychological tests conducted at this time also showed increased prevalence of maladaptive personality traits[49] and cognitive rigidity, slow processing, and impaired cognitive control.[51] The long-lasting differences identified between the two study groups underscored outright the lack of brain plasticity and the irreversible nature of the nutritional insult despite comprehensive intervention.

In addition to the increased behavioral and cognitive impairments in the affected Barbados Nutrition Study cohort, there were also identifiable differences in social outcome between malnourished and control participants. Of note were the poorer socioeconomic outcomes observable among the previously malnourished individuals.[52] This was demonstrated through the substantially widening gap in weekly household income between the index and control groups over their life spans, likely mediated by differences in behavioral and cognitive functioning. We also speculated that the reduced socioeconomic circumstances of these individuals could very likely impact the cognitive development of their children, thereby perpetuating the effects of the original insult into the next generation.

In the most recent phase of the Barbados work, we have opted to assess functioning in the offspring of the previously malnourished individuals and of the healthy comparison group. Initial results indicated significant transgenerational effects on attention and cognitive performance. Importantly, differences on measures of attention, IQ, and executive control were similar to those seen in the original cohort, even though this generation had never personally experienced any episodes of malnutrition. Parallel epigenetic studies of the G1 and G2 participants confirmed altered gene expression (S. Akbarian *et al.*, unpublished data). In these preliminary epigenetic analyses, we found significant correlations between DNA methylation levels with metabolic indexes and various metrics for adverse cognitive and behavioral outcomes. These results are currently being validated.

4. INTERGENERATIONAL STUDIES OF MALNUTRITION: ANIMAL STUDIES

Animal studies have provided the clearest evidence of intergenerational effects of malnutrition on brain and behavioral outcomes. Although these findings cannot be directly extrapolated to the human,[41,53] they provide a clear indication that long-term research on malnutrition in human populations is necessary. Early animal studies by Cowley and Griesel[54] over three generations and by Chow[55] over two generations demonstrated that the nutritional status of the rat mother had effects on the growth and development of subsequent generations of rats. Later, Bresler *et al.*[56] and Zamenhof *et al.*[57] lengthened the time frame of their study and also expanded the research criteria to include behavioral parameters. In studies that extended to 10 generations of both maternal malnutrition and paternal malnutrition, these investigators demonstrated increasing growth and behavioral deficits over each subsequent generation. In a parallel series of studies, Stewart *et al.*[58,59] similarly reported a pattern of increasing growth deficits involving both body length and weight in rats with up to 12 generations of moderate protein malnutrition (8% protein diet vs. 25% for controls). Subsequently, they also documented that, after exposure to 10–12 generations on the low-protein diet, three generations of rehabilitation on an adequate protein diet were required for the rats to achieve the same growth rate as control rats, but performance on a Lashley jumping stand did not improve.[60] When we were given the original Stewart colony in 1973, in the meantime transferring the rats from England to Cambridge, MA, the research focus was broadened to include comprehensive studies of behavioral and cognitive outcomes following intergenerational malnutrition.[61–72]

In the ensuing series of reports, malnutrition exposure was extended to 20 generations, which was then followed by a period of nutritional rehabilitation that sought to reverse the effects of the nutritional deficits within one to four generations.[63,64,66] The rats were divided into two groups and were provided with either a low-protein diet (7.5% casein) or an adequate level of protein (25% casein), with supplementation of methionine added to both diets. A third group was derived through cross-fostering a select number of rats from the low-protein, intergenerationally malnourished group to healthy control mothers; these rats were then provided with adequate nutrition after weaning (25% casein) and studied over the next few generations. We found that deficits in mother–pup interaction[68–70] persisted up to four generations after instituting dietary rehabilitation. In the studies of the pups during their early development, we identified significant deficits persisting to four generations of nutritional rehabilitation on a test of home-orienting behavior, which allows pups to find the maternal nest when displaced from it.[63,72] Further, we confirmed that, in adulthood, rats with histories of intergenerational malnutrition were significantly more impaired on learning tests than animals with a single generation of malnutrition.[64,67] Following dietary rehabilitation for up to two generations, we documented recovery on the Lashley jumping stand test among the females but observed no such recovery in the males.[64]

Studies of growth and development in this colony showed that rats given a low-protein diet for multiple generations were stunted in size, but physical growth relative to controls was quickly recovered after the administration of an adequate protein diet. In fact, after one generation on the adequate (25%) protein diet, we found that physical growth exceeded growth rates observed in the control rats.[69] Further, in the first-generation rehabilitated animals, we found that they exhibited increased reproductive incompetence in adulthood as compared with controls.[73] Thus, while physical size recovered quickly after dietary rehabilitation, reproductive competence and certain behavioral and cognitive functions were impaired, even after several generations. Similar to the Barker hypothesis,[12] our data suggested the presence of biological adaptations to the intergenerational history of malnutrition, changes that persisted long after an adequate diet was provided. Because these experimental studies were carried out many years ago, however, mechanisms underlying the transgenerational effects we observed, including possible epigenetic mechanisms, were not studied and remain unknown.

Finally, it is interesting to note that findings exhibited in a series of animal studies of intergenerational malnutrition on insulin production are parallel to our own findings on the long-term effects of malnutrition in both rats and humans. Reusens *et al.* and Pinheiro *et al.* looked at long-term and transgenerational effects of maternal malnutrition in the rat through analysis of the endocrine pancreas and associated transmission of altered glucose metabolism.[74,75] Aets and Van Assche found that prenatal malnutrition was linked with diabetogenic effects in the adult offspring, which resultantly impacted their pregnancies and the next generation.[76] Another study examining the inheritance of an insulin-resistant phenotype also demonstrated the transgenerational presence of abnormal glucose metabolism, similarly supporting the conclusion of nutritionally induced heritable mechanisms.[77] Thus, prenatal malnutrition and postnatal malnutrition in the rat have been implicated in an increased risk of transgenerational diabetes.

Relatedly, in both U.S. and Belgian studies, experimental diabetes, produced by injecting streptozotocin (a beta cell killer) *in utero*, has been associated with increased prevalence of diabetes over several generations.[78,79] These studies indicate that the offspring of treated females ultimately developed diabetes, in turn producing offspring with a greater likelihood of developing diabetes than the offspring of nondiabetic mothers. The offspring of the second-generation pregnancies also had abnormal glucose tolerance tests.

The relevance of the findings of these animal studies may also lie in their direct relationship to results previously procured in human populations, such as that of the Pima Indians, whose diabetes appears to be generationally transmitted,[80] and other high-risk populations with prior exposure to malnutrition. These populations, including that of the Dutch famine study,[81] the Chinese famine study,[82] and the Barbados Nutrition Study, have at this point already demonstrated an increased rate of diabetes within a transgenerational context.

In summary, despite the paucity of published human and animal studies relating to intergenerational malnutrition, the consistency of the existing evidence strongly supports the intergenerational effects of malnutrition on brain and behavior. Likely mechanisms underlying these effects, which require further investigation, can include environmentally induced epigenetic changes that often result in a compromised intrauterine environment and altered programming and gene expression. In addition, a history of malnutrition is also associated with aberrant patterns of child rearing and fewer educational and professional opportunities, factors that may both impact future generations.

5. INTERGENERATIONAL STUDIES OF EARLY ADVERSITY AND TRAUMA: HUMAN AND ANIMAL STUDIES

Environmental factors including early life adversity and trauma exposure may also play an important role in contributing to the risk of psychological disorders in adulthood and transgenerationally. As is the case in studies of malnutrition, the impact of early childhood trauma is similarly thought to be mediated in large part by epigenetic mechanisms that can alter gene expression and perpetuate lifelong effects into subsequent generations.[83,84]

To this end, a recent study by Yehuda *et al.* identified an increased risk of PTSD and mood disorders, including anxiety disorders and, to a less extent, substance abuse disorders, in the offspring of Holocaust survivors.[85,86] These investigators found that maternal PTSD was specifically associated with increased reports of PTSD in the adult offspring. Moreover, relative to controls, these adult offspring reported having higher levels of childhood trauma through self-administered surveys, especially trauma categorized as emotional abuse and neglect. The extent of perceived childhood abuse was significantly correlated with the severity of PTSD reported by these subjects. In addition, the degree of childhood emotional abuse was found to be precisely associated with 24-h urinary cortisol levels. This study suggested that the experience of childhood trauma may represent an important link in the transmission of PTSD from parent to child.

Although Yehuda *et al.*[85] looked predominantly at environmental factors that may play a role in the transgenerational transmission of PTSD, other studies have confirmed the importance of maternal care relevant to such transmission. A recently conducted study involving a population of children whose parents were exposed to the Rwandan genocide serves as one such example.[87] This study suggested that child-rearing practices were most closely aligned with the resultant onset of a child's PTSD and that the increased familial risk could not be explained by genetics alone or by other factors such as parenting constitutional factors, namely, temperament. Thus, although constitutional factors may play an important role, the authors of this report highlighted the importance of the rearing environment for young children and their mental health. Interestingly, another recent study demonstrates explicitly that perceived maternal care in childhood can modify fetal growth of the next-generation offspring, of which brain development is the major part. Neuwald *et al.*[88] reported that fetal growth interacts with perceived maternal care and affects attentional skills of the offspring at 18 months of age. The findings of both of these studies further support the role of environment as a modifier of trauma transmission. Epigenetic changes may provide an explanation as to how such transmission occurs.

Relatedly, an epigenetics-centered follow-up of Yehuda *et al.*'s[85] PTSD cohort found the previously identified transgenerational outcomes associated with trauma to be linked to changes in the glucocorticoid receptor (GR).[89] Adult offspring of Holocaust survivor mothers with PTSD had higher levels of GR-1 methylation than controls. However, offspring with two parents with a history of PTSD demonstrated lower levels of GR-1 methylation, a finding that was also associated with greater poststeroid cortisol suppression. A study by McGowan *et al.* similarly implicated abuse in the early environment in the establishment of altered methylation patterns in the brain, especially that of the GR.[90] Methylation in the neuron-specific GR promoter in the brains of suicide victims who experienced childhood abuse was significantly higher than those measured in suicide victims who were not abused, or in controls. Increased methylation was correlated with decreased GR expression in these brains.[90]

The demonstrated relationship in humans between mental health disorders and childhood maltreatment has led researchers to investigate the potential role of changes in DNA methylation using animal models. In the rat, variations in the amount of maternal care provided during the litter period have been shown to impact the pattern of DNA methylation in the pup's brain at adult ages. Based on naturally occurring individual differences in maternal care provided by female rats, Meaney and Szyf,[91] McGowan *et al.*,[92] and Zhang *et al.*[93] explored epigenetic mechanisms that may underlie such variations. Female rats were selected based on their nurturing ability and placed in two experimental groups, high arched back nursing and licking/grooming (ABN–LG) and low ABN–LG. Studies of pups reared by high ABN–LG mothers showed that these pups, when adult, had higher GR expression in the hippocampus and stronger glucocorticoid feedback sensitivity than pups reared by low ABN–LG mothers.[94] The pups with high ABN–LG rearing showed lower levels of DNA methylation in the promoter region of the GR and an associated increase in the binding of the nerve growth factor-inducible protein-A transcription factor to the promoter region of GR.[94] In a follow-up study, Weaver *et al.* reported that all effects of high ABN–LG rearing could be reversed in the adult offspring by intracerebroventricular infusion of L-methionine, which acts as a methyl group donor.[95] The results obtained from the later study suggest that certain epigenetic changes established during the early postnatal period of development of the rat can be reversed in adulthood, thus emphasizing the plasticity of the adult rat brain with regard to the epigenome.

In another study of early life maltreatment, rat pups received daily abusive maternal care from nonbiological mothers in the first week of life (postnatal days 1–7), which included stepping, dropping, dragging, and active avoidance.[96] The maltreated pups in this study were reported as having reduced BDNF expression and increased BDNF methylation in the prefrontal cortex as adults. The maternal abuse during early development led to poor mothering in the exposed female offspring when they reared their own young; cross-fostering to healthy mothers did not fully reverse these changes in methylation levels.[96] Altered BDNF methylation in the brain of the offspring from the maltreated mothers may have resulted from (1) transmission of methylation via the gametes or (2) changes in methylation elicited by the transmission of poor maternal care across generations.

These findings are noteworthy because, in humans, increased methylation of the BDNF gene in the frontal cortex is associated with major psychoses, including schizophrenia and bipolar disorder.[97] Thus, maltreatment during early development may predispose not only individuals to adverse mental health outcomes but also their offspring to similar conditions via transmission of abnormal methylation patterns across generations.

6. OTHER INTERGENERATIONAL STUDIES: CHEMICAL EXPOSURES

Recent evidence of intergenerational effects of preconception drug abuse has been summarized in a review by Vassoler *et al.*[98] Parental abuse of drugs, even without direct fetal exposure, may impact the physiology and behavior of the offspring and later generations. Given the difficulty in separating out human abuse from other environmental and biological factors associated with addiction, including malnutrition, most available studies to date have made use of animal models, and epigenetic mechanisms are thought to underlie these effects.[99]

Alcohol, the most common drug of abuse worldwide, has been reported as producing changes in basal activity levels in both humans[100] and rodents.[101] Additionally, documented behavioral effects include impaired learning, memory, and attention in both human studies[102] and rodent models[103,104] and also increased anxiety and depression-like phenotypes in rodents.[105,106]

In the human, there is limited evidence of intergenerational transmission relating to both maternally and paternally linked alcohol abuse.[107] A recent three-generation study reported an increased risk of alcohol abuse in the G3 offspring.[108] Jamerson *et al.* identified that paternal alcohol exposure of rats that ended prior to fertilization resulted in neurobehavioral changes in the progeny.[109] Animal studies have also confirmed transgenerational effects of other drugs of abuse in nonexposed offspring, including those of parental[110] opioid[111] and cocaine[112] use.

Because most of the reported human and animal studies of drug abuse have only been conducted across two generations, further generations need to be studied to confirm epigenetic mechanisms impacted by such insults. Analyses of studies that have preliminarily looked beyond two generations seem to suggest that the behavioral and cognitive phenotypes persist. The specific mechanisms underlying the generational transmission of alcohol and drug abuse remain unknown, however, and may be associated with underlying epigenetic phenomena, a likely possibility given the evidence of patrilineal inheritance patterns accompanying most of these exposures.

7. INTERGENERATIONAL TRANSMISSION: EPIGENETIC MECHANISMS

Recent advances in science and technology have underscored the plausibility of attaining precise biological mechanisms to explain the transgenerational transmission of the effects of various early environmental insults, and there is already evidence that points to the ability of environmental factors, especially those of nutrition and trauma, to alter chromatin state. For example, a current report directly links nutritional intake and epigenetic transmission to the next generation in a group of 18–45-year-old Gambian women.[113] In this study, analysis of 12 plasma biomarkers confirmed that seasonal variations modified the intake of specific methyl-donor groups, especially those associated with vitamin B6 and folate. The level of these biomarkers predicted increased or decreased methylation at metastable epialleles in DNA extracted postnatally from lymphocytes and hair follicles of their infants. The findings from this study demonstrated that maternal nutritional status in the periconception period resulted in persistent and systemic epigenetic changes in human metastable epialleles that are passed on to the next generation. The functional consequences of these changes, however, remain unknown. As such, transgenerational epigenetic inheritance is the best-candidate mechanism to explain such transgenerational effects, especially those associated with male-line transmission, which have now been reported in a number of studies relating to both malnutrition[31] and drug abuse.[114]

The potential role of epigenetic mechanisms has also been noted in a number of papers documenting evidence of changes in DNA methylation many years after exposure to malnutrition, including the 60-year follow-up study of the prenatally malnourished offspring of women exposed to the Dutch famine.[115] When compared with same-sex siblings who had no exposure to prenatal malnutrition, the exposed study participants demonstrated a reduction in DNA methylation of the imprinted IGF2 gene, a gene with an important role in determining growth and development. This association was specific for periconceptional exposure, reinforcing the conclusion that early stages of development are critical for establishing and maintaining epigenetic marks. In another study by the same group, lower methylation levels of the INSIGF gene and higher levels of methylation in *IL-10*, *LEP*, *ABCA1*, *GNASAS*, and *MEG3* genes were also reported in adulthood.[116] These genes are known to play a role in inflammatory and metabolic processes, as well as cardiovascular disease.

In animal studies of maternal malnutrition, there is already confirmed evidence that methylation changes may occur in the context of reduced maternal dietary intake at levels that do not necessarily impact birth weight. For example, a 30% reduction in maternal intake altered gene expression and methylation levels in monkey kidneys in the absence of any demonstrable reduction in birth weight.[117,118] Functional consequences of such "hidden forms" of malnutrition in the rat have previously been reported as capable of producing brain and behavioral deficits without impacting size at birth or postnatal growth rates.[119] Hence, use of birth and postnatal size as a proxy for perinatal health status is inadequate and should be supplemented by more comprehensive nutritional measurements.

Although numerous human and animal studies have identified changes in gene expression, the mechanisms underlying transgenerational transmission are not yet well understood. The intergenerational reappearance of acquired epigenetic marks, which was once deemed impossible, has now been documented in both plants and animals.[114] It was previously thought that the process of epigenome erasure and resetting during fertilization excluded any possibility of intergenerational transmission of acquired epigenetic marks. This conclusion evolved from early studies showing that global DNA methylation levels were much lower after fertilization when compared with methylation levels present in mature gametes and after implantation. While it is true that most of the acquired marks are, in fact, likely erased early after conception, the emphasis of current research is now on how and why certain acquired marks survive classical intergenerational reprogramming.

In fact, it has been suggested that demethylation of the genome following fertilization may in fact be nonexistent. This conclusion is attributed largely to the work of Li and O'Neill who observed instead that such demethylation in mice appears to be the result of antigenic masking. Through the employment of a variety of different testing methods, these researchers demonstrated that methylation was maintained after fertilization and throughout early embryo development.[120] Franklin and Mansuy similarly demonstrated epigenetic modifications of the genome in a mouse model, successfully tracing such changes through several generations.[121]

Relevantly, alterations in DNA methylation of several genes have now been observed in sperm of both F2 and F3 males of early stressed mice.[122] Radford *et al.* recently published a high-profile transgenerational study in a mouse model of malnutrition that was the first to report explicit alterations of the germline DNA methylome.[123] In this model, malnutrition was induced through restriction of the maternal diet in the last week of pregnancy.[124] While decreased methylation in the Radford study was not maintained in adult males in the next generation, locus-specific gene expression was nevertheless perturbed. Hence, these studies strongly suggest that acquired modifications in behavior and brain function due to malnutrition and other environmental stressors are transmitted to later generations through the germ line.

Finally, because environmental influences can alter the epigenetic profile not only of the child-bearing female but also of her developing fetus, it has been proposed that changes in the second generation are not truly transgenerational. Graff and Mansuy[125] have therefore suggested that even in animal models, only effects documented in the third generation should truly be considered as intergenerational effects.[125,126] Further epidemiological and basic research is consequently needed to document intergenerational changes in preclinical and human populations, in order to identify specific patterns of epigenetic change and their explicit relationship to brain and behavioral functions. Future studies of this nature are also necessary to foster the development of realistic approaches to reversing environmental effects on programming.

8. CONCLUSION

Early insults during critical periods of brain development, both prenatal and postnatal, can result in epigenetic changes that may impact health and behavioral outcomes over the life span and into future generations. There is ample evidence that these early stages of brain development are sensitive to various environmental insults, including malnutrition, childhood trauma, and drug exposures. The notion that such changes, both physiological and behavioral, can also carry over into subsequent generations has long been recognized, especially in the context of experimental studies. However, epigenetic mechanisms capable of explaining such phenomena were not available until relatively recently, with most of this research published only within the last decade.

Although researchers have clearly demonstrated chromatin modifications in animal studies involving early environmental stressors and their long-term effects on behavior and brain function, there are relatively few intergenerational studies involving human populations. As such, future long-term research is needed in order to more precisely document the impact of chromatin remolding on changes in the brain and the intergenerational transmission of these effects.

ACKNOWLEDGMENTS

This research was supported by grants (to J. G.) from the National Institutes of Health (HD060986 and MH 074811). The authors would like to thank Jacqueline S. DeAnda for her assistance in preparing this manuscript. We also acknowledge the contributions of the late Dr. Peter Morgane and the late Sir Dr. Frank C. Ramsey, who were integral to our research program on the long-term effects of childhood malnutrition.

REFERENCES

[1] Grantham-McGregor S, Cheung YB, Cueto S, Glewwe P, Richter L, Strupp B. Developmental potential in the first 5 years for children in developing countries. *Lancet*. 2007;369(9555):60–70.

[2] Prado EL, Dewey KG. Nutrition and brain development in early life. *Nutr Rev*. 2014;72(4):267–284.

[3] Mussey RD. Nutrition and human reproduction; an historical review. *Am J Obstet Gynecol*. 1949;57(6):1037–1048.

[4] Baird D. The contribution of obstetrical factors to serious physical and mental handicap in children. *J Obstet Gynaecol Br Emp*. 1959;66:743–747.

[5] Thomson A. Maternal stature and reproductive efficiency. *Eugen Rev*. 1959;51(3):157.

[6] Thomson AM, Billewicz WZ. Nutritional status, maternal physique and reproductive efficiency. *Proc Nutr Soc*. 1963;22:55–60.

[7] Birch HG, Gussow JD. *Disadvantaged Children: Health, Nutrition and School Failure.* New York, NY: Harcourt, Brace & World; 1970.
[8] Birch HG, Richardson SA, Baird D, Horobin G, Illsley R. *Mental Subnormality in the Community: A Clinical and Epidemiological Study.* Baltimore, MD: Williams and Wilkins; 1970.
[9] Ounstead M, Scott A, Ounstead C. Transmissions through the female line of a mechanism constraining human fetal growth. *Ann Hum Biol.* 1986;13:143–151.
[10] Hackman E, Emanuel I, van Belle G, Daling J. Maternal birth weight and subsequent pregnancy outcome. *JAMA.* 1983;250(15):2016–2019.
[11] Emanuel I, Filakti H, Alberman E, Evans SJ. Intergenerational studies of human birthweight from the 1958 birth cohort. 1. Evidence for a multigenerational effect. *Br J Obstet Gynaecol.* 1992;99(1):67–74.
[12] Barker DJP. *Mothers, Babies and Health in Later Life.* Edinburgh: Churchill Livingstone: A Division of Harcourt, Brace and Company Limited; 1998.
[13] Lumey LH. Decreased birthweights in infants after maternal in utero exposure to the Dutch famine of 1944–1945. *Paediatr Perinat Epidemiol.* 1992;6(2):240–253.
[14] Horta BL, Gigante DP, Osmond C, Barros FC, Victora CG. Intergenerational effect of weight gain in childhood on offspring birthweight. *Int J Epidemiol.* 2009;38(3): 724–732.
[15] Emanuel I. Maternal health during childhood and later reproductive performance. *Ann NY Acad Sci.* 1986;477:27–39.
[16] Barker DJ, Godfrey KM, Gluckman PD, Harding JE, Owens JA, Robinson JS. Fetal nutrition and cardiovascular disease in adult life. *Lancet.* 1993;341(8850):938–941.
[17] Nathanielsz PW, Nathanielsz P. *Life in the Womb: The Origin of Health and Disease.* Ithaca, NY: Promethean Press; 1999.
[18] Barker DJ. Coronary heart disease: a disorder of growth. *Horm Res.* 2003;59(suppl 1): 35–41.
[19] Barker DJ, Bull AR, Osmond C, Simmonds SJ. Fetal and placental size and risk of hypertension in adult life. *BMJ.* 1990;301(6746):259–262.
[20] Forsen T, Eriksson JG, Tuomilehto J, Teramo K, Osmond C, Barker DJ. Mother's weight in pregnancy and coronary heart disease in a cohort of Finnish men: follow up study. *Br Med J.* 1997;315(7112):837–840.
[21] Leon DA, Koupilova I, Lithell HO, et al. Failure to realise growth potential in utero and adult obesity in relation to blood pressure in 50 year old Swedish men. *BMJ.* 1996;312(7028):401–406.
[22] Rich-Edwards JW, Stampfer MJ, Manson JE, et al. Birth weight and risk of cardiovascular disease in a cohort of women followed up since 1976. *BMJ.* 1997;315(7105):396–400.
[23] Ravelli AC, van der Meulen JH, Michels RP, et al. Glucose tolerance in adults after prenatal exposure to famine. *Lancet.* 1998;351(9097):173–177.
[24] Stanner SA, Bulmer K, Andres C, et al. Does malnutrition in utero determine diabetes and coronary heart disease in adulthood? Results from the Leningrad siege study, a cross sectional study. *BMJ.* 1997;315(7119):1342–1348.
[25] Salonen MK, Kajantie E, Osmond C, et al. Role of childhood growth on the risk of metabolic syndrome in obese men and women. *Diabetes Metab.* 2009;35(2): 94–100.
[26] Barker DJ, Forsen T, Uutela A, Osmond C, Eriksson JG. Size at birth and resilience to effects of poor living conditions in adult life: longitudinal study. *BMJ.* 2001;323(7324): 1273–1276.
[27] Kajantie E, Osmond C, Barker DJ, Forsen T, Phillips DI, Eriksson JG. Size at birth as a predictor of mortality in adulthood: a follow-up of 350 000 person-years. *Int J Epidemiol.* 2005;34(3):655–663.

[28] Lahti M, Raikkonen K, Lemola S, et al. Trajectories of physical growth and personality dimensions of the Five-Factor Model. *J Pers Soc Psychol.* 2013;105(1):154–169.
[29] Lahti M, Raikkonen K, Wahlbeck K, et al. Growth in infancy and childhood and hospitalization for personality disorders in adulthood: the Helsinki Birth Cohort Study. *J Pers Disord.* 2011;25(5):620–633.
[30] Raikkonen K, Forsen T, Henriksson M, et al. Growth trajectories and intellectual abilities in young adulthood: the Helsinki Birth Cohort study. *Am J Epidemiol.* 2009;170(4):447–455.
[31] Kaati G, Bygren LO, Pembrey M, Sjostrom M. Transgenerational response to nutrition, early life circumstances and longevity. *Eur J Hum Genet.* 2007;15(7): 784–790.
[32] Painter RC, Osmond C, Gluckman P, Hanson M, Phillips DI, Roseboom TJ. Transgenerational effects of prenatal exposure to the Dutch famine on neonatal adiposity and health in later life. *BJOG.* 2008;115(10):1243–1249.
[33] Veenendaal MV, Painter RC, de Rooij SR, et al. Transgenerational effects of prenatal exposure to the 1944–45 Dutch famine. *BJOG.* 2013;120(5):548–553.
[34] Lumey LH, Stein AD. Offspring birth weights after maternal intrauterine undernutrition: a comparison within sibships. *Am J Epidemiol.* 1997;146(10):810–819.
[35] Stein AD, Barnhart HX, Hickey M, Ramakrishnan U, Schroeder DG, Martorell R. Prospective study of protein-energy supplementation early in life and of growth in the subsequent generation in Guatemala. *Am J Clin Nutr.* 2003;78(1):162–167.
[36] Behrman JR, Calderon MC, Preston SH, Hoddinott J, Martorell R, Stein AD. Nutritional supplementation in girls influences the growth of their children: prospective study in Guatemala. *Am J Clin Nutr.* 2009;90(5):1372–1379.
[37] Walker SP, Wachs TD, Gardner JM, et al. Child development: risk factors for adverse outcomes in developing countries. *Lancet.* 2007;369(9556):145–157.
[38] Venables PH, Raine A. Poor nutrition at age 3 and schizotypal personality at age 23: the mediating role of age 11 cognitive functioning. *Am J Psychiatr.* 2012;169(8): 822–830.
[39] Dobbing J. Vulnerable periods of brain development. In: *lipids, Malnutrition and the Developing Brain. Ciba Foundation Symposium*; 1971:9–29.
[40] Morgane PJ, Austin-LaFrance R, Bronzino J, et al. Prenatal malnutrition and development of the brain. *Neurosci Biobehav Rev.* 1993;17(1):91–128.
[41] Galler JR, Shumsky JS, Morgane PJ. Malnutrition and brain development. In: Walker WA, Watkins JB, eds. *Nutrition in Pediatrics: Basic Science and Clinical Application.* Neuilly-sur-Seine, France: J. B. Decker Europe, Inc.; 1996:196–212
[42] Galler JR, et al. Long-term effects of early kwashiorkor compared with marasmus. I. Physical growth and sexual maturation. *J Pediatr Gastroenterol Nutr.* 1987c;6(6):841–846.
[43] Galler JR, et al. The influence of early malnutrition on subsequent behavioral development. II. Classroom behavior. *J Am Acad Child Adolesc Psychiatr.* 1983;22(1):16–22.
[44] Galler JR, et al. The long-term effects of early kwashiorkor compared with marasmus. IV. Performance on the national high school entrance examination. *Pediatr Res.* 1990;28(3):235–239.
[45] Galler JR, et al. Infant malnutrition is associated with persisting attention deficits in middle adulthood. *J Nutr.* 2012b;142(4):788–794.
[46] Galler JR, et al. Long-term effects of early kwashiorkor compared with marasmus. II. Intellectual performance. *J Pediatr Gastroenterol Nutr.* 1987a;6(6):847–854.
[47] Galler JR, Bryce CP, Waber D, et al. Early childhood malnutrition predicts depressive symptoms at ages 11–17. *J Child Psychol Psychiatry.* 2010;51(7):789–798.
[48] Waber DP, Eaglesfield D, Fitzmaurice GM, Bryce C, Harrison RH, Galler JR. Cognitive impairment as a mediator in the developmental pathway from infant

malnutrition to adolescent depressive symptoms in Barbadian youth. *J Dev Behav Pediatr.* 2011;32(3):225–232.
[49] Galler JR, Bryce CP, Zichlin ML, et al. Malnutrition in the first year of life and personality at age 40. *J Child Psychol Psychiatry.* 2013;54(8):911–919.
[50] Waber DP, et al. Impaired IQ and academic skills in adults who experienced moderate to severe infantile malnutrition: a 40-year study. *Nutr Neurosci.* 2014b;17(2):58–64.
[51] Waber DP, Bryce CP, Fitzmaurice GM, et al. Neuropsychological outcomes at midlife following moderate to severe malnutrition in infancy. *Neuropsychology.* 2014;28(4):530–540.
[52] Galler JR, et al. Socioeconomic outcomes in adults malnourished in the first year of life: a 40-year study. *Pediatrics.* 2012a;130(1):e1–e7.
[53] Galler J, Kanis K. Animal models of malnutrition applied to brain research. In: Rassin D, Harber B, Drujan B, eds. *Basic and Clinical Aspects of Nutrition and Brain Development.* New York, NY: Alan R Liss; 1987:57–73.
[54] Cowley JJ, Griesel RD. The effect on growth and behaviour of rehabilitating first and second generation low protein rats. *Anim Behav.* 1966;14(4):506–517.
[55] Chow BF. Effect of maternal dietary protein on anthropometric and behavioral development of the offspring. *Adv Exp Med Biol.* 1974;49:183–219.
[56] Bresler DE, Ellison G, Zamenhof S. Learning deficits in rats with malnourished grandmothers. *Dev Psychobiol.* 1975;8:315–323.
[57] Zamenhof S, van Marthens E. Effects of prenatal and chronic undernutrition on aging and survival in rats. *J Nutr.* 1982;112(5):972–977.
[58] Stewart RJ, Preece RF, Sheppard HG. Recovery from long-term protein-energy deficiency. *Proc Nutr Soc.* 1973;32(3):103a.
[59] Stewart RJ, Preece RF, Sheppard HG. Twelve generations of marginal protein deficiency. *Br J Nutr.* 1975;33(2):233–253.
[60] Stewart RJ, Sheppard H, Preece R, Waterlow JC. The effect of rehabilitation at different stages of development of rats marginally malnourished for ten to twelve generations. *Br J Nutr.* 1980;43(3):403–412.
[61] Galler JR. Home orientation in nursling rats: the effects of rehabilitation following intergenerational malnutrition. *Dev Psychobiol.* 1979;12(5):499–508.
[62] Galler JR, Rosenthal M. The effects of intergenerational malnutrition on the mother—pup relationship in rats. *Infant Behav Dev.* 1979;2:1–12.
[63] Galler JR. The effects of intergenerational and postnatal malnutrition on the behavioral development of rats. In: Brozek J, ed. *Behavioral Effects of Energy and Protein Defecits.* 1979b:22–38.
[64] Galler JR. Home-orienting behavior in rat pups surviving postnatal or intergenerational malnutrition. *Dev Psychobiol.* 1980;13(6):563–572.
[65] Galler JR, Manes M. Gender differences in visual discrimination by rats in response to malnutrition of varying durations. *Dev Psychobiol.* 1980;13(4):409–416.
[66] Galler JR, et al. Varying deficits in visual discrimination performance associated with different forms of malnutrition in rats. *J Nutr.* 1980;110(2):231–240.
[67] Galler JR, Seelig C. Home-orienting behavior in rat pups: the effect of 2 and 3 generations of rehabilitation following intergenerational malnutrition. *Dev Psychobiol.* 1981;14(6):541–548.
[68] Galler JR, Propert KJ. Maternal behavior following rehabilitation of rats with intergenerational malnutrition. 2. Contribution of mothers and pups to deficits in lactation-related behaviors. *J Nutr.* 1981a;111(8):1337–1342.
[69] Galler JR, Propert KJ. Maternal behavior following rehabilitation of rats with intergenerational malnutrition. 1. Persistent changes in lactation-related behaviors. *J Nutr.* 1981b;111(8):1330–1336.

[70] Galler JR. Visual discrimination in rats: the effects of rehabilitation following intergenerational malnutrition. *Dev Psychobiol.* 1981;14(3):229–236.
[71] Galler JR, Propert KJ. Early maternal behaviors predictive of the survival of suckling rats with intergenerational malnutrition. *J Nutr.* 1982;112(2):332–337.
[72] Galler JR. Intergenerational effects of undernutrition. In: Kavanagh JF, ed. *Understanding Mental Retardation: Research Accomplishments and New Frontiers*. Baltimore, MD: Paul H. Brooks Publishing Company; 1988:35–40.
[73] Galler JR, Zartarian G. Reproductive performance in rats with different histories of malnutrition. *Br J Nutr.* 1981;45(2):251–255.
[74] Reusens B, Remacle C. Programming of the endocrine pancreas by the early nutritional environment. *Int J Biochem Cell Biol.* 2006;38(5–6):913–922.
[75] Pinheiro AR, Salvucci ID, Aguila MB, Mandarim-de-Lacerda CA. Protein restriction during gestation and/or lactation causes adverse transgenerational effects on biometry and glucose metabolism in F1 and F2 progenies of rats. *Clin Sci (London).* 2008;114(5):381–392.
[76] Aets L, Van Assche FA. Animal evidence for the transgenerational development of diabetes mellitus. *Int J Biochem Cell Biol.* 2006;38:894–903.
[77] Thamotharan M, Garg M, Oak S, et al. Transgenerational inheritance of the insulin-resistant phenotype in embryo-transferred intrauterine growth-restricted adult female rat offspring. *Am J Physiol Endocrinol Metab.* 2007;292(5):E1270–E1279.
[78] Van Assche F, Aerts L. Long-term effect of diabetes and pregnancy in the rat. *Diabetes.* 1985;34(suppl 2):116–118.
[79] Oh W, Gelardi NL, Cha CJ. The cross-generation effect of neonatal macrosomia in rat pups of streptozotocin-induced diabetes. *Pediatr Res.* 1991;29(6):606–610.
[80] Knowler WC, Bennett PH, Hamman RF, Miller M. Diabetes incidence and prevalence in Pima Indians: a 19-fold greater incidence than in Rochester, Minnesota. *Am J Epidemiol.* 1978;108(6):497–505.
[81] van Abeelen AF, Elias SG, Bossuyt PM, et al. Famine exposure in the young and the risk of type 2 diabetes in adulthood. *Diabetes.* 2012;61(9):2255–2260.
[82] Zheng X, Wang Y, Ren W, et al. Risk of metabolic syndrome in adults exposed to the great Chinese famine during the fetal life and early childhood. *Eur J Clin Nutr.* 2012;66(2):231–236.
[83] Gudsnuk K, Champagne FA. Epigenetic influence of stress and the social environment. *ILAR J.* 2012;53(3–4):279–288.
[84] Kundakovic M, Champagne FA. Early-life experience, epigenetics, and the developing brain. *Neuropsychopharmacology*. 2014 [EPub ahead of print].
[85] Yehuda R, Bell A, Bierer LM, Schmeidler J. Maternal, not paternal, PTSD is related to increased risk for PTSD in offspring of Holocaust survivors. *J Psychiatr Res.* 2008;42(13):1104–1111.
[86] Yehuda R, Halligan SL, Grossman R. Childhood trauma and risk for PTSD: relationship to intergenerational effects of trauma, parental PTSD, and cortisol excretion. *Dev Psychopathol.* 2001;13(3):733–753.
[87] Roth M, Neuner F, Elbert T. Transgenerational consequences of PTSD: risk factors for the mental health of children whose mothers have been exposed to the Rwandan genocide. *Int J Ment Health Syst.* 2014;8(1):12.
[88] Neuwald MF, Agranonik M, Portella AK, et al. Transgenerational effects of maternal care interact with fetal growth and influence attention skills at 18 months of age. *Early Hum Dev.* 2014;90(5):241–246.
[89] Yehuda R, Daskalakis NP, Lehrner A, et al. Influences of maternal and paternal PTSD on epigenetic regulation of the glucocorticoid receptor gene in Holocaust survivor offspring. *Am J Psychiatry*. 2014;171(8):872–880.

[90] McGowan PO, Meaney MJ, Szyf M. Diet and the epigenetic (re)programming of phenotypic differences in behavior. *Brain Res.* 2008;1237:12–24.
[91] Meaney MJ, Szyf M. Environmental programming of stress responses through DNA methylation: life at the interface between a dynamic environment and a fixed genome. *Dialogues Clin Neurosci.* 2005;7(2):103–123.
[92] McGowan PO, Suderman M, Sasaki A, et al. Broad epigenetic signature of maternal care in the brain of adult rats. *PLoS One.* 2011;6(2):e14739.
[93] Zhang TY, Labonte B, Wen XL, Turecki G, Meaney MJ. Epigenetic mechanisms for the early environmental regulation of hippocampal glucocorticoid receptor gene expression in rodents and humans. *Neuropsychopharmacology.* 2013;38(1):111–123.
[94] Weaver IC, Cervoni N, Champagne FA, et al. Epigenetic programming by maternal behavior. *Nat Neurosci.* 2004;7(8):847–854.
[95] Weaver IC, Champagne FA, Brown SE, et al. Reversal of maternal programming of stress responses in adult offspring through methyl supplementation: altering epigenetic marking later in life. *J Neurosci.* 2005;25(47):11045–11054.
[96] Roth TL, Lubin FD, Funk AJ, Sweatt JD. Lasting epigenetic influence of early-life adversity on the BDNF gene. *Biol Psychiatry.* 2009;65(9):760–769.
[97] Mill J, Tang T, Kaminsky Z, et al. Epigenomic profiling reveals DNA-methylation changes associated with major psychosis. *Am J Hum Genet.* 2008;82(3):696–711.
[98] Vassoler FM, Byrnes EM, Pierce RC. The impact of exposure to addictive drugs on future generations: physiological and behavioral effects. *Neuropharmacology.* 2014;76(pt B):269–275.
[99] Vassoler FM, White SL, Schmidt HD, Sadri-Vakili G, Pierce RC. Epigenetic inheritance of a cocaine-resistance phenotype. *Nat Neurosci.* 2013;16(1):42–47.
[100] Goodwin DW, Schulsinger F, Hermansen L, Guze SB, Winokur G. Alcoholism and the hyperactive child syndrome. *J Nerv Ment Dis.* 1975;160(5):349–353.
[101] Abel EL. Paternal alcohol exposure and hyperactivity in rat offspring: effects of amphetamine. *Neurotoxicol Teratol.* 1993;15(6):445–449.
[102] Tarter RE, Jacob T, Bremer DL. Specific cognitive impairment in sons of early onset alcoholics. *Alcohol Clin Exp Res.* 1989;13(6):786–789.
[103] Abel EL. Effects of physostigmine on male offspring sired by alcohol-treated fathers. *Alcohol Clin Exp Res.* 1994;18(3):648–652.
[104] Abel EL, Lee JA. Paternal alcohol exposure affects offspring behavior but not body or organ weights in mice. *Alcohol Clin Exp Res.* 1988;12(3):349–355.
[105] Abel EL. Alcohol consumption does not affect fathers but does affect their offspring in the forced swimming test. *Pharmacol Toxicol.* 1991;68(1):68–69.
[106] Abel EL. Paternal alcohol consumption affects grooming response in rat offspring. *Alcohol.* 1991;8(1):21–23.
[107] Schmidt CM, Tauchmann H. Heterogeneity in the intergenerational transmission of alcohol consumption: a quantile regression approach. *J Health Econ.* 2011;30(1):33–42.
[108] Handley ED, Chassin L. Alcohol-specific parenting as a mechanism of parental drinking and alcohol use disorder risk on adolescent alcohol use onset. *J Stud Alcohol Drugs.* 2013;74(5):684–693.
[109] Jamerson PA, Wulser MJ, Kimler BF. Neurobehavioral effects in rat pups whose sires were exposed to alcohol. *Brain Res Dev Brain Res.* 2004;149(2):103–111.
[110] Byrnes JJ, Johnson NL, Schenk ME, Byrnes EM. Cannabinoid exposure in adolescent female rats induces transgenerational effects on morphine conditioned place preference in male offspring. *J Psychopharmacol.* 2012;26(10):1348–1354.
[111] Byrnes JJ, Babb JA, Scanlan VF, Byrnes EM. Adolescent opioid exposure in female rats: transgenerational effects on morphine analgesia and anxiety-like behavior in adult offspring. *Behav Brain Res.* 2011;218(1):200–205.

[112] He F, Lidow IA, Lidow MS. Consequences of paternal cocaine exposure in mice. *Neurotoxicol Teratol.* 2006;28(2):198–209.
[113] Dominguez-Salas P, Moore SE, Baker MS, et al. Maternal nutrition at conception modulates DNA methylation of human metastable epialleles. *Nat Commun.* 2014;5:3746.
[114] Grossniklaus U, Kelly WG, Ferguson-Smith AC, Pembrey M, Lindquist S. Transgenerational epigenetic inheritance: how important is it? *Nat Rev Genet.* 2013;14(3):228–235.
[115] Heijmans BT, Tobi EW, Stein AD, et al. Persistent epigenetic differences associated with prenatal exposure to famine in humans. *Proc Natl Acad Sci USA.* 2008;105(44):17046–17049.
[116] Tobi EW, Lumey LH, Talens RP, et al. DNA methylation differences after exposure to prenatal famine are common and timing- and sex-specific. *Hum Mol Genet.* 2009;18(21):4046–4053.
[117] Unterberger A, Szyf M, Nathanielsz PW, Cox LA. Organ and gestational age effects of maternal nutrient restriction on global methylation in fetal baboons. *J Med Primatol.* 2009;38(4):219–227.
[118] Nijland MJ, Mitsuya K, Li C, et al. Epigenetic modification of fetal baboon hepatic phosphoenolpyruvate carboxykinase following exposure to moderately reduced nutrient availability. *J Physiol.* 2010;588(pt 8):1349–1359.
[119] Resnick O, Morgane PJ, Hasson R, Miller M. Overt and hidden forms of chronic malnutrition in the rat and their relevance to man. *Neurosci Biobehav Rev.* 1982;6(1):55–75.
[120] Li Y, O'Neill C. Persistence of cytosine methylation of DNA following fertilisation in the mouse. *PLoS One.* 2012;7(1):e30687.
[121] Franklin TB, Mansuy IM. Epigenetic inheritance in mammals: evidence for the impact of adverse environmental effects. *Neurobiol Dis.* 2010;39(1):61–65.
[122] Franklin TB, Russig H, Weiss IC, et al. Epigenetic transmission of the impact of early stress across generations. *Biol Psychiatry.* 2010;68(5):408–415.
[123] Radford EJ, Ito M, Shi H, et al. In utero undernourishment perturbs the adult sperm methylome and intergenerational metabolism. *Science.* 2014;345(6198):1255903.
[124] Jimenez-Chillaron JC, Hernandez-Valencia M, Reamer C, et al. Beta-cell secretory dysfunction in the pathogenesis of low birth weight-associated diabetes: a murine model. *Diabetes.* 2005;54(3):702–711.
[125] Graff J, Mansuy IM. Epigenetic codes in cognition and behaviour. *Behav Brain Res.* 2008;192(1):70–87.
[126] Jablonka E, Raz G. Transgenerational epigenetic inheritance: prevalence, mechanisms, and implications for the study of heredity and evolution. *Q Rev Biol.* 2009;84(2):131–176.

CHAPTER EIGHT

The Future of Neuroepigenetics in the Human Brain

Amanda Mitchell[*,†], **Panos Roussos**[*,†,‡], **Cyril Peter**[*,†], **Nadejda Tsankova**[†,§], **Schahram Akbarian**[*,†]

[*]Department of Psychiatry, Icahn School of Medicine at Mount Sinai, New York, USA
[†]Department of Neuroscience, Icahn School of Medicine at Mount Sinai, New York, USA
[‡]Department of Genetics and Genomic Sciences, Icahn School of Medicine at Mount Sinai, New York, USA
[§]Department of Pathology, Icahn School of Medicine at Mount Sinai, New York, USA

Contents

Abstract

Complex mechanisms shape the genome of brain cells into transcriptional units, clusters of condensed chromatin, and many other features that distinguish between various cell types and developmental stages sharing the same genetic material. Only a few years ago, the field's focus was almost entirely on a single mark, CpG methylation; the emerging complexity of neuronal and glial epigenomes now includes multiple types of DNA cytosine methylation, more than 100 residue-specific posttranslational histone modifications and histone variants, all of which superimposed by a dynamic and highly regulated three-dimensional organization of the chromosomal material inside the cell nucleus. Here, we provide an update on the most innovative approaches in neuroepigenetics and their potential contributions to approach cognitive functions and disorders unique to human. We propose that comprehensive, cell type-specific mappings

Progress in Molecular Biology and Translational Science, Volume 128
ISSN 1877-1173
http://dx.doi.org/10.1016/B978-0-12-800977-2.00008-5

of DNA and histone modifications, chromatin-associated RNAs, and chromosomal "loopings" and other determinants of three-dimensional genome organization will critically advance insight into the pathophysiology of the disease. For example, superimposing the epigenetic landscapes of neuronal and glial genomes onto genetic maps for complex disorders, ranging from Alzheimer's disease to schizophrenia, could provide important clues about neurological function for some of the risk-associated noncoding sequences in the human genome.

1. INTRODUCTION

1.1. Chromatin and epigenetic regulation: General principles

The elementary unit of chromatin is the nucleosome, or 146 bp of genomic DNA wrapped around an octamer of core histones, connected by linker DNA and linker histones. As further described below, the collective set of DNA and histone modifications and variant histones provide the molecular substrates of the epigenome, here broadly defined as the epigenetic landscapes that define the functional architecture of the chromosomal material, including transcriptional and many other features of genome organization that are differentially regulated in different cell types and developmental stages of the organism.[1,2]

1.1.1 DNA (hydroxy)methylation

Two related but functionally very different types of DNA modifications, cytosine C5-methylation (5mC) and hydroxymethylation (5hmC) of cytosines in CpG dinucleotides, provide the bulk of the epigenetic modifications in vertebrate DNA.[3] There are additional types of DNA modifications, including 5-formylcytosine (5fC) and 5-carboxylcytosine (5caC), which are viewed as chemical breakdown products, from mC5 to hmC5 to 5fC to 5caC, and in addition may carry regulatory functions.[4,5] While the majority of DNA (hydroxy)methylation is found at sites of CpG dinucleotides and, more generally, in the CpG-enriched sequences of the genome, a much larger fraction, or up to 25% of mC5, is found at non-CpG sites in the brain.[6] The mC5 and hmC5 markings show a differential pattern of genomic occupancy, with the hmC5 mark concentrated toward the 5′ end of the genes and the proximal most portion of transcriptional units, broadly correlating with local gene expression levels,[7–9] and a potential role in the regulation of intron/exon boundaries and splicing events of neuron-specific gene transcripts.[10] On the other hand, less than 3% of methylcytosine (mC5) markings are positioned around the 5′ end of the genes.[11]

1.1.2 Histone modifications

There is evidence that far more than 100 amino acid residue-specific posttranslational modifications (PTMs) exist in the vertebrate cell,[12] including monomethylation (me1), dimethylation (me2), and trimethylation (me3); acetylation and crotonylation; poly-ADP-ribosylation; and small protein (ubiquitin, SUMO) modification of specific lysine residues, as well as arginine (R) methylation and citrullination, serine (S) phosphorylation, tyrosine (T) hydroxylation, and several others.[12–14] These site- and residue-specific PTMs often define chromatin structure and function, with an epigenetic histone code (a combinatorial set of histone PTMs) differentiating between promoters, gene bodies, enhancer and other regulatory sequences, and condensed heterochromatin.[15] It is important to emphasize that histone PTMs rarely occur in isolation and instead multiple histone PTMs appear to be coregulated and, as a group, define the aforementioned chromatin states.[16] Many active promoters, for example, are defined by high levels of histone H3 lysine 4 trimethylation in combination with various histone lysine acetylation markings, while repressive histone PTMs, including the trimethylated forms of H3K9, H3K27, and H4K20, potentially colocalize to some of the same loci in the genome and so forth.[15] Proteins associated with the regulation of histone PTMs are sometimes referred to as "writers" or "erasers" or "readers," essentially differentiating the process of establishing or removing a mark as opposed to its docking functions for chromatin remodeling complexes that regulate transcription or induce and maintain chromatin condensation.[14,17,18]

1.1.3 Histone variants, chromatin remodeling, and nucleosome positioning

In addition to the core histones H2A/H2B/H3/H4, a number of histone variants, with H3.3, H3.1, H3.2, H2A.Z, and H2A.X, are some of the best-studied examples. Variant histones, which differ from the canonical histone at few amino acid positions, could affect nucleosome stability and compaction.[19]

Chromatin remodeling complexes are composed of multiple subunits that, according to their classical definition, regulate sliding and mobility of nucleosomes, powered by ATP hydrolysis, thereby regulating gene expression and RNA polymerase II access at transcription start sites.[20]

Examples of well-known chromatin remodelers with a critical role in brain development include the BAF (SWI/SNF) complex and CHD family of proteins.[20] Mutations in numerous members of the BAF complex and multiple CHD proteins have now been linked to psychiatric disease and developmental brain disorders, with additional investigations in mutant mice.[20–22]

1.1.4 Chromatin-associated RNAs

While the process of gene expression is obviously defined by nascent RNA emerging from genomic DNA packaged into chromatin, chromatin-associated RNAs (caRNAs) (or cbRNA for chromatin-bound RNA) could be reserved to RNA species as part of a chromatin structure, thereby regulating its functions. According to some estimates, up to 2–3% of the nucleic acid content in chromatin is contributed by polyadenylated RNAs.[23] One of the best-known examples of a caRNA is provided by the X-chromosome inactive transcript that governs the silencing of the second X in female diploid cells.[24,25] Some caRNAs, including a complex long noncoding RNA, composed of 148 exons and introns, at a neurodevelopmental risk locus on chromosome *15q11–13*,[26] produce "RNA clouds" in *cis*, thereby triggering lasting decondensation of the surrounding chromosomal material.[27,28] In addition, some of the caRNAs expressed in the human brain, including a noncoding RNA at the *DPP10* (chromosome 2q14.1) locus, show human-specific epigenetic regulation and could contribute to cognitive features and disease vulnerabilities not shared with other primate species.[29]

1.1.5 Higher order chromatin

DNA methylation, epigenetic decoration of nucleosomal (including variant) histones, and the various caRNAs/cbRNAs described to date still would fall short to adequately describe the epigenome and localized chromatin architectures at any given genomic locus. This is because the chromosomal arrangements in the interphase nucleus are not random. Specifically, loci at sites of active gene expression are more likely to be clustered together and situated toward a central position within the nucleus, while heterochromatin and silenced loci move more toward the nuclear periphery.[30,31] Chromosomal loopings, in particular, are among the most highly regulated supranucleosomal structures and are associated with transcriptional regulation, by, for example, positioning distal regulatory enhancer or silencer elements that—in the linear genome—are positioned potentially many hundred kilobases apart from a gene, to interact directly with that specific promoter.[32,33] Proper regulation of such types of higher order chromatin is certainly of critical importance for orderly brain development and function. For example, Cornelia de Lange syndrome (CdLS) with an estimated incidence of 1:10–30,000 live births among the more frequent genetic disorders (source: http://ghr.nlm.nih.gov) is associated with severe developmental delay and a range of neuropsychiatric symptoms.[34] CdLS (including *Online Mendelian Inheritance in Man* (*OMIM*) 122470 and 300590) involves causative mutations in the cohesin complex, a multisubunit protein that includes, among others, nipped-B-like protein (NIPBL), structural maintenance of chromosomes 1A and 3 (SMC1A and SMC3) proteins, and histone deacetylase 8 (HDAC8).[35,36] Cohesin is thought to form ring-like structures bringing together DNA segments from different locations, and by interaction with transcriptional coactivators, the complex could promote the physical interaction of promoters with enhancers separated by thousands of base pairs on the linear genome, thereby regulating cell type-specific gene expression programs.[37]

2. CHALLENGES FOR EPIGENETIC APPROACHES IN THE (HUMAN) BRAIN

2.1. Cellular specificity of epigenetic markings

Conventional chromatin assays designed to detect and quantify DNA methylation and histone modifications require an input material between 10^3 and 10^8 nuclei.[38,39] Such types of assay typically lack cellular resolution, which poses a challenge given that brain tissue is composed of an extremely heterogeneous mixture of different cell types, including glia-to-neuron ratios that could show considerable fluctuations across normal development or in certain disease states, such as conditions associated with neurodegeneration or neuroinflammation. To date, many studies exploring epigenetic dysregulation of gene expression in major psychiatric disease examined DNA methylation and histone modifications in tissue homogenates, while typically, the gene(s) of interest often is expressed only in a select subpopulation of neurons or other cells.[40–48] On the other hand, there is evidence that each cell type is defined by differential regulation of DNA methylation and histone modifications at hundreds or thousands of promoter and enhancer sequences, resulting in considerable "epigenetic distance" even between cortical neurons and their surrounding glia and other nonneuronal cells.[49–51] Of note, statistical methodology has been developed to decompose neuronal and nonneuronal signals in DNA methylation assays on tissue homogenate.[52,53] These *in silico* approaches could be extremely valuable not only for already existing datasets but also for future studies in which sorting is not feasible, including in projects in which the cell type-specific nuclear epitope is not reliably expressed in every postmortem specimen. In these cases, cell type-specific epigenomic maps could be constructed in a smaller set of brains, while the larger cohort undergoes deconvolution based on epigenomic maps in tissue homogenate.

Given the inherent problem of cellular heterogeneity in the brain, many groups have turned to fluorescence-activated cell sorting (FACS) for the isolation of more homogenous cell populations, including neurons, microglia, oligodendrocytes, astrocytes, and neural/glial progenitors at various stages of their lineage differentiation. Several useful markers exist to define these cell types *in situ*, but only few of them have shown the ability to label live cells for isolation (Table 8.1). A major obstacle to using most phenotypic markers for FACS is the necessity to target an extracellular cell-surface domain, particularly if downstream purification is needed.[58,62] Several surface markers have been used to isolate oligodendrocyte progenitor cells (OPCs) A2B5,[57] NG2,[56] and CD140a-PDGFRα[55]; neural stem cells (NSCs) CD133[60]; microglia CD11b[61]; and most recently astrocyte-lineage cells from mice GLAST[58] and from expanded human NSCs CD44[59] (Table 8.1). While some of the abovementioned markers have sorted populations that are still quite heterogeneous,[56,57,60] others have shown significant enrichment for specific cell types.[55,58,61] The use of cell-surface antibodies to isolate cells is not ideal: cytoplasmic processes of more mature glia are often partially destroyed during enzymatic digestion and the rapidly flowing stream of FACS, leading to significant decrease in final cell yield. Cell sorting with the surface markers CD44 and GLAST (ASCA-1 antibody) has allowed the isolation of adult glia from postmortem subventricular zone tissue, providing sufficient yield of high-quality DNA for downstream genomic analysis (1000–100,000 cell events depending on cell population, brain age, and postmortem time) but insufficient for chromatin immunoprecipitation analysis (Tsankova lab, unpublished data). Furthermore, the use of cell-surface antibodies for FACS necessitates fresh tissue, due to the inevitable cell lysis of many cells during freezing and thawing. This adds another constraint: fresh autopsy tissue with short postmortem time or surgical brain tissue is not readily available and easily accessible outside of large neurosurgical centers or close collaboration with neuropathologists and neurosurgeons.

The limitation of using surface markers has caused many to turn to transgenic strategies for FACS in rodents, allowing the use of cytoplasmic or

Table 8.1 Common markers defining cell-type identity/lineage specificity of brain cells

Marker	Cell type/lineage	Cellular localization	FACS fresh human tissue	Technique	Cell specificity
NeuN	Neuron	Nucleus	Y: Nuclei	Nuclei extraction > NeuN + FACS	Mature neurons
PDGFRα	Oligodendrocyte-lineage	Extracellular cell surface	Y: Whole cell	Papain dissociation > CD140a + FACS	OPCs
NG2	Oligodendrocyte-lineage +	Cell surface	Y: Whole cell (hESCs)	hESCs > NG2 + FACS	OPCs, pericytes, endothelium
A2B5	Oligodendrocyte-lineage +	Cell surface	Y: Whole cell	Papain dissociation > A2B5 + MACS	OPCs, astrocyte-lineage cells, immature neurons
Olig2	Oligodendrocyte-lineage +	Nucleus	N		OPCs to mature oligodendrocytes, few multipotent glial progenitors
GLAST	Astrocyte-lineage	Extracellular cell surface	Y: Whole cell (mice)	Trypsin dissociation > ACSA-1 + MACS	Astrocyte-lineage cells
CD44	Astrocyte-lineage +	Extracellular cell surface	Y: Whole cell (hNSC)	Expanded hNSC > CD44 + CD184 + FACS	Fibrous astrocytes (SVZ, WM), glial progenitors, inflammatory cells
GFAP	Astrocyte-lineage	Cytoplasm	N		Astrocytes, some neural stem cells/ radial glia
ALDH1L1	Astrocyte-lineage	Cytoplasm	N		Astrocyte-lineage cells

Continued

Table 8.1 Common markers defining cell-type identity/lineage specificity of brain cells—cont'd

Marker	Cell type/lineage	Cellular localization	FACS fresh human tissue	Technique	Cell specificity
CD133	Neural stem cell +	Extracellular cell surface	Y: Whole cell	Enzymatic dissociation > CD133 + CD34 – CD45 – FACS	Neural stem cells, ependyma, endothelium
Sox2	Neural stem cell	Nucleus	N		Embryonic and adult neural stem cells
Nestin	Neural stem cell +	Cytoplasm	N		Neural stem cells, ependyma
CD11b	Microglia	Extracellular cell surface	Y: Whole cell	Enzymatic dissociation > CD11b + CD45dim MACS	Microglia, other monocytes

Some of these markers have allowed isolation of cell populations using fresh human brain tissue by FACS, such as neurons (via NeuN[54]), oligodendrocyte-lineage cells (via PDGFRα,[55] NG2,[56] and A2B5[57]), astrocyte lineage cells (via GLAST,[58] in mice and CD44[59] in expanded human NSCs), neural stem cells (via CD133[60]), and microglia (via CD11b[61]). +, more than one cell population isolated by FACS; MACS, magnetic-activated cell sorting; SVZ, subventricular zone; WM, white matter.

nuclear markers with good cell-type specificity, coexpressed with a fluorescent reporter; one example is the use of GFP–GFAP reporter mice for isolation of astrocytes.[63] This approach, unfortunately, is not applicable for human tissue. An alternative, successful strategy to collect nuclei from human tissue has been the isolation of neuronal nuclei labeled with the specific antibody NeuN.[64] This overcomes the caveats of heterogeneity, fragility of cytoplasmic processes, and availability of fresh tissue and offers satisfactory yield for downstream epigenetic studies. Protocols are available for efficient purification and immunotagging of nuclei from frozen postmortem brain specimens, which then could be processed by fluorescence-activated sorting. Thus, the separate collection of 10^7–10^8 neuronal and nonneuronal nuclei from a few hundred milligram of postmortem cerebral cortex in a single day is possible, thereby enabling separate processing of neuronal and nonneuronal chromatin.[54,65,66] A similar neuronal marker for astrocytes has not yet been described, but the oligodendrocyte-lineage nuclear marker Olig2 could offer a comparable nuclear isolation solution for immature and maturing oligodendrocytes.[66] Another exciting avenue to explore is the use of molecular beacons as mRNA-specific targets for FACS in human brain tissue, which has already shown feasibility in rodents.[62]

2.2. With focus on the candidate gene approach, only few high-resolution epigenomic mappings

With the exception of neurosurgical cases, the bulk of human brain studies, including those focused on neuropsychiatric disorders, rely on postmortem tissue. There is general consensus that nucleosomal arrays, histone modifications, and the activity of histone-modifying enzymes such as methyl- and acetyltransferases, DNA methylation markings, and even some of the chromosomal loop formations are preserved, at least partially, in postmortem brain tissue that typically is exposed to 5–30 or more hours of autolysis time before being stored in a −70 °C freezer.[38,64,67–69] Thus, 20 years ago, epigenetic exploration of the diseased human brain started out with restriction enzyme-based DNA cytosine methylation mapping at a predetermined set of CpG dinucleotides surrounding the 5′ end of amyloid-beta precursor (*APP*)[70] and *FMR1* genes in single cases diagnosed with Alzheimer's disease and fragile X mental retardation syndrome,[71–73] followed 10 years later by histone methylation mappings at the site of NMDA glutamate receptor genes in developing cerebral and cerebellar cortices.[74] In the case of fragile X, the expansion of CGG codon from (normally) 5 to 40 repeats to 50 to over 200 repeats triggers excessive DNA methylation at the promoter, effectively shutting down gene expression by silencing the surrounding chromatin.[75] These highly reproducible molecular phenotypes in a monogenetic neurodevelopmental disorder then provided a road map for similar studies on disorders of a more heterogeneous and complex etiology, including Alzheimer's dementia, depression, schizophrenia, and others.[40–42,44,48,70] However, small cohort sizes, typically involving far less than 50–100 brains, in conjunction with the overall only very subtle differences between cases and controls[45,46,48] and the considerable degree of sample heterogeneity in terms of etiology of disease, have thus far precluded a general consensus on which, if any, genomic locus shows a reproducible epigenetic defect in any of the aforementioned neuropsychiatric disorders.

In a very different line of research, large, multi-institutional consortia, such as the *Encyclopedia of DNA Elements* (*ENCODE*), have harnessed powerful, next-generation sequencing-based technologies[76,77] to generate comprehensive genome-scale maps, often at (near) base-pair resolution, for DNA methylation and histone modification landscapes, chromatin accessibility as measured by DNase I hypersensitivity, and transcription factor binding profiles in a variety of peripheral cell lines and tissues.[78,79] In contrast, only very few epigenetic markings, including DNA methylation and hydroxymethylation,[80] and a small number of histone methylation and acetylation markings that differentiate between active and inactive/repressed promoter and enhancer sequences[81] have been charted with next-generation sequencing technology in the human brain and with a narrow focus mostly on epigenetic changes that occur during normal development of the cerebral cortex. Furthermore, with few exceptions,[82] such types of studies were based on very small and even single-digit cohorts.[81] In the nearby future, government-sponsored program such as *PsychENCODE* (www.grants.gov) and/or private- or industry-sponsored efforts should catalyze the generation of a much larger brain epigenomic dataset than the one currently available, with the expectation of a deeper understanding about the epigenetic mechanisms of normal development and aging and changes in chronic neuropsychiatric disease.

2.3. Higher order chromatin studies in the human brain

Despite the growing realization of the importance of chromosomal looping and other higher order chromatin structures for transcriptional regulation (discussed above), very little is known about their role in the nervous system, including only a handful of studies with human brain tissue.[29,83,84] This is surprising given that chromosome conformation capture (commonly referred to as *3C*[85,86]) as the standard approach to map chromosomal loopings, is applicable to brain tissue collected postmortem, at least for some the genomic loci that were examined so far.[83] The 3C technique explores physical interactions between DNA fragments separated by Kb or Mb of interspersed sequence; cross-linked chromatin is digested with a specific restriction enzyme, religated and amplified using primer pairs for which forward and reverse primers match to different portions of the genomic locus of interest.[83]

One interesting example of 3C applied on human brain involves *GAD1*, encoding the 67Kda glutamic acid decarboxylase GABA synthesis enzyme.[84] A loop, initially detected in the prefrontal cortex and then verified in neuronal cultures derived from pluripotent skin cells, connected sequences surrounding the *GAD1* transcription start site with a region 50 kb further upstream. This loop showed a significant weakening in the prefrontal cortex of some subjects with schizophrenia that were affected by decreased *GAD1* expression. There was evidence that the loop was conserved between the rodent and primate brain, and indeed, based on 3C studies in reporter mice, this loop appeared to be much stronger in cortical GABAergic interneurons (which express *Gad1*) compared with other cortical cells (which do not express *Gad1*). Such type of study could be viewed as "proof of principle" that higher order chromatin is (i) amenable to analyses in the human brain, (ii) potentially altered in common brain disorders, and (iii) potentially amenable to translational approaches and follow-up work in preclinical model systems, including human cell culture and animals.

Similar to DNA methylation and histone modification studies discussed above, there is a glaring vacuum of genome-scale and agnostic (not candidate gene-based) higher order chromatin studies in the human brain. This is not due to the lack of basic techniques. At the time this review was written, massively parallel sequencing has enabled the production of chromosome conformation libraries to detect interactions (i) between specific loci genome-wide using circular chromosome conformation capture (4C), (ii) between a large set of loci using carbon-copy chromosome conformation capture (5C), (iii) between all regions in the genome using HiC, and (iv) between all regions in the genome bound by a specific protein or histone mark using chromatin immunoprecipitation for using paired-end sequencing (ChIA-PET).[87–90] The datasets that are grounded in these techniques have provided unique insights into the organizational complexities and nonrandomness of spatial genome architectures, with multiple types of looping interaction groups between highly transcribed genes, gene-rich active regions, nonactive centromere proximal clusters, coregulated genes, gene ontology groups, and so on.[91–93] We predict that for the human brain too, a similar or perhaps even more complex organization will emerge, once neuroscientists and neuropsychiatrists embark on these types of studies.

HiC is the most high-throughput method, investigating interactions across the entire genome with a resolution of 5–10 kb using the best current algorithms starting with 300 million mappable reads or four lanes of Illumina sequencing.[94,95] This technique, though powerful, is not always the most amendable method to detect higher order chromatin organization due to the depth of sequencing required. 4C, however, can be used to determine genome-wide interaction contacts for specific loci with a 1–8 kb resolution using only 500,000 mappable reads, allowing for the multiplexing of several experiments in one lane of Illumina sequencing.[96,97] Both HiC and 4C are limited, however, due to the sheer number of interactions that exist at neighboring loci, which tend to result in a large majority of interaction read counts. 5C can overcome neighboring interaction artifacts from high-throughput sequencing by only designing neighboring primers on opposite strands and primers on the same strand apart by 15–30 kb. Thus, the determination of which "C" to use for individual experiments depends on the hypothesis in question.

3. EVIDENCE AND DEBATE

3.1. Epigenetic markings in the brain: State or trait?

The rationale of exploring certain types of epigenetic modifications in postmortem brain of subjects diagnosed with psychiatric disease is, as mentioned above, often based on the hypotheses that changes in RNA expression are associated with altered epigenetic decoration at the site of the corresponding gene promoter and related regulatory sequences. Quite often, the accompanying abnormalities in DNA methylation and histone modifications are then interpreted in terms of a stable and long-lasting epigenetic "lesion" in response to an environmental insult or some other pathogenic effect operating in early life, many years before the brain was obtained at autopsy. For example, different grades of maternal care in the early postnatal period lead to differential regulation of promoter-associated DNA methylation and histone acetylation at the aforementioned disease gene, *Gad1*, in the hippocampus of adult rats,[98] and likewise, deficits in open chromatin-associated histone methylation at *GAD1* in the prefrontal cortex of adult schizophrenics was discussed in the context of defective neurodevelopment in conjunction with a risk haplotype at the promoter.[43] Furthermore, based on postmortem studies in adult suicide victims, there is evidence that the suffering of abuse in early childhood years leaves a lasting DNA methylation imprint on the stress-regulated glucocorticoid receptor *NR3C1* promoter[99] in the hippocampus and on DNA repeats encoding ribosomal RNAs.[100] In addition, normal aging may be associated with widespread age-related changes in gene expression in the cerebral cortex, including downregulation of many neuronal genes.[48,101–103]

However, it is fair to admit that little is known about the stability and dynamic turnover of epigenetic markings in human brain, and therefore, it remains unclear whether any of the aforementioned epigenetic alterations in the brain of adult psychiatric subjects indeed reflect a (mal)adaptive "trait" stably maintained for years or, alternatively, whether disease-associated chromatin changes merely reflect the brain's functional state at the time of death. The "trait" hypothesis appears very plausible in the context of monogenetic disorders associated with aberrant and excessive repressive DNA and histone methylation in *cis* (at the site of the mutation). Examples include the aforementioned *CGG* triplet expansion at the *FMR1* (fragile X) gene promoter[75] or the *GAA* triplet repeat expansion in the first intron of the *FRATAXIN* gene associated with Friedreich's ataxia, an autosomal recessive neurodegenerative condition.[104] In these cases, the epigenetic dysregulation is firmly linked to the pathophysiology of disease (resulting from silenced gene expression), and there can be little doubt that the observed changes in (postmortem) brain chromatin, like the impairments in neurological functions, most likely existed across the entire life span.[73,104] Furthermore, there are many other examples strongly suggesting that the DNA sequence variation is a major driver for epigenetic differences between subjects. In addition to the abovementioned example gene, *GAD1*, many other single-nucleotide polymorphisms (SNPs) across the entire genome, including those that have been genetically implicated in the risk of major psychiatric disease (including bipolar disorder and schizophrenia), exhibit a robust effect on methylcytosine levels at the site of nearby genes.[103,105] However, studies in mono- and dizygotic twins and related work in animals convincingly demonstrated that molecular mechanisms of heritability are unlikely due to DNA sequence differences alone.[106]

On the other hand, it will be difficult to confirm whether many of the reported epigenetic alterations observed in small cohorts of "sporadic" cases with schizophrenia, autism, depression, and other psychiatric diseases represent a type of molecular alteration stably related to the underlying disease. Given that most or perhaps all epigenetic markings studied to date are subject to bidirectional regulation in the cell culture system and animal model, it is reasonable to assume that the epigenetic decoration of human brain genomes is subject to similar types of dynamic regulation.

For example, DNA methylation at specific promoter sequences is subject to rapid up- or downregulation on the scale of minutes to hours.[107,108] Hippocampal DNA methylation signatures are highly sensitive to acute depolarization[109,110] and electroconvulsive seizures, affecting regulatory sequences regulating NMDA and GABA-A receptor genes, Notch signaling pathways, and other systems with a key regulatory role for synaptic signal and plasticity.[111]

Furthermore, changes in neuronal activity result in robust changes in expression and activity of multiple DNA methylation-associated proteins with an essential role for neuronal health and function, including the methyl-CpG-binding protein MeCP2 or Gadd45b that recruits cytidine deaminases and thymidine glycosylases at genomic sites subject to active DNA methylation.[112–114] Furthermore, physiological activation of hippocampal circuitry during learning and memory is sufficient to elicit highly dynamic DNA methylation changes at *PP1*, *REELIN*, and other gene promoters regulating synaptic plasticity.[115,116] The complex molecular machineries mediating demethylation of CpG dinucleotides or histone lysine residues are becoming increasingly understood, which is a remarkable progress given that these and other types of epigenetic modifications were not long ago considered to be potentially irreversible.[117–120] These findings, taken together, would suggest that some of the epigenetic alterations reported in diseased postmortem brain are not necessarily stable for very long periods of time and, instead, are regulated by mechanisms that operate on a much shorter timescale, perhaps lasting only a few weeks or days or even less. Clinically relevant conditions that reportedly affect chromatin structure and function in the brain include ischemia,[121] exposure to environmental toxins,[122–124] abuse of nicotine,[125,126] alcohol,[127] psychostimulants,[128–130] and antipsychotic and mood-stabilizing drugs.[45,131–136]

3.2. Mapping brain epigenomes from the culture dish?

While a significant portion of the epigenomic organization in brain nuclei is preserved in postmortem specimens stored at brain banks (with autolysis time typically in the range of 5–35 h), animal models suggest that some histone modification types and chromosomal loopings and other higher order chromatin structures show significant signal decay and other secondary effects already after 12–15 h of tissue autolysis and decay.[38,68] Because access to fresh, neurosurgically obtained brain tissue is not an option for the vast majority of subjects diagnosed with neuropsychiatric disease, it is not surprising that many investigators in the field are harnessing pluripotent stem cells (iPS) from the skin or other cell types as the starting material to model brain tissue in the culture dish.

Once the generation of pluripotent stem cells by reprogramming somatic cells via retroviral transduction of four transcription factors (i.e., Oct4, Sox2, Klf4, and c-Myc)[137] had been accomplished, the technique has been further advanced[138] and applied by multiple groups to generate iPS-derived neuronal and glial cultures or even a three-dimensional organoid mimicking the cerebral cortex in a dish.[139] These include neurodevelopmental syndromes[140,141] and schizophrenia.[142] Such type of studies undoubtedly will pave the way for broader future iPS-based approaches that most certainly will become a mainstay in the field of biological psychiatry. Cellular reprogramming with subsequent neural differentiation, including neural circuitry and active synapse formation in the dish, is likely to mimic many key steps of neurodevelopment and opens up the possibility of conducting electrophysiological recordings and other functional assays on nervous tissues of living subjects.[142] Therefore, iPS technology provides an unprecedented opportunity to study the molecular and cellular biology of the nervous system from any patient (or at least from those who are able to give consent). A subset of psychiatric susceptibility genes may even, as in case of the *MYLT1* transcription factor, which when mutated confers high risk for neurodevelopmental disability, promote the process of neuronal differentiation from stem cell preparations *ex vivo*.[143,144]

These recent advances in reprogramming technologies have also fueled general interest in the field to explore epigenetic regulation of the nervous system in patient- and control-derived iPS. For example, several studies explored chromatin structures and synaptic signaling in neuronal cultures of Rett syndrome (RTT) patients with *MECP2* mutations and controls.[141,145] The *MECP2* gene product, methyl-CpG-binding protein 2, is highly expressed in the nervous system and occupies widespread territories of neuronal chromatin, dependent on the local density of methyl-CpG-dinucleotides.[146] Loss-of-function mutations and other *MECP2* structural variants have been linked to RTT, a disorder of early childhood associated with developmental and cognitive regression and a broad range of neurological symptoms.[147,148] Furthermore, work on reprogrammed skin cells of Rett patients, which then was further confirmed in the *Mecp2* mutant mouse brain, uncovered a genomic instability phenotype defined by disinhibition and increased mobility of retrotransposon and other parasitic DNA element activities due to altered DNA and histone methylation, in conjunction with changes in the global chromatin state.[145,146] This work illustrates the promising potential of neural cultures, derived from skin fibroblasts, to study epigenetic (dys)regulation in specific disease cases and to gain knowledge about the molecular underpinnings of neurological disorders. Furthermore, the short-chain fatty acid derivative and anticonvulsant and mood-stabilizing drug valproic acid (VPA) induces pluripotency from skin fibroblasts when coadministered with Oct-4 and Sox-2 transcription factors.[149] Furthermore, VPA promotes neuronal differentiation from progenitor stages[150,151] and the drug induces, in cell culture and the brain, the upregulation of open chromatin-associated histone acetylation and methylation markings at promoters of genes with a key role in neurotransmission.[43,152,153] It will be interesting to quantify these VPA-dependent effects on pluripotency and neuronal differentiation and to compare VPA treatment responders to nonresponders. More broadly, pharmacoepigenomics, or a drug's direct and indirect effects on chromatin structure and function, may perhaps in the future emerge as an interesting biomarker to predict treatment response and side effects or illuminate novel, hitherto unsuspected mechanisms of drug action.

However, despite all these unprecedented perspectives of generating nervous tissue in the dish from skin cells, it is important to point out that the technique still faces hurdles and challenges, particularly in the context of epigenetic regulation. This is because the eraser and subsequent redecoration of epigenetic markings across the genome—a key mechanism for successful reprogramming—may be incomplete at some loci, resulting in carryover effects so that the reprogrammed cells (iPS) maintain DNA methylation signatures that define the original donor cell type (e.g., fibroblasts).[154,155] Furthermore, while some of the chromosomal loopings and higher order chromatin structures that regulate neuronal gene expression in the human brain are tractable in stem cell-derived neuronal cultures, others are not.[83,84]

Given the considerable variability of epigenetic and cellular phenotypes after reprogramming,[156,157] it will be difficult to faithfully "rebuild" in the culture dish the cortical neuronal networks with their constituents such as pyramidal neurons and their surrounding inhibitory cells. This is a challenge for any disease-related study, including many psychiatric disorders that are likely to show only subtle differences in cellular response patterns, as compared with controls. While purely speculative at this point in time, imagine the potential benefits of the iPS technology for pharmaco(epi)genomics and treatment response paradigm. For example, iPS-derived cells could be "challenged" with a compound and the epigenome signature measured to distinguish between treatment-resistant and treatment-responsive patients. In this context, one interesting biomarker appears to relate to histone methylation levels and acetylation levels in the peripheral blood cells of subjects exposed to the histone deacetylase inhibitor, valproate.[158]

3.3. Functional neuroepigenomics to inform disease-associated variants

Over the last decade, multiple genome-wide association (GWA) studies have produced strongly significant evidence that specific common DNA genetic variants among people influence their genetic susceptibility to a number of complex neuropsychiatric illnesses, including schizophrenia[159] and bipolar disorder.[160] The majority of common variant loci associated with genetic risk for these complex diseases reside within noncoding sequence of unknown function, and many are far from discovered genes. To mention just one example, consider the major histocompatibility complex (MHC) locus that has long been implicated in psychiatric disease, including three large genome-wide association studies (GWAS) published jointly in 2009.[161–163] These studies identified up to 45 disease-associated SNPs in the 26–33 megabase region of the MHC locus on chromosome 6, but strikingly, fifty percent of these SNPs were not located near genes. Some of the SNPs with the statistically strongest disease association were approximately 30 kb away from the nearest gene.

Furthermore, the disease-associated genomic regions are frequently large and often contained multiple implicated genetic variants due to local linkage disequilibrium patterns. In order to be able to understand these associations mechanistically, it is critical to develop strategies for honing in on regions and genetic variants more likely to have functional effects. Thus, the elucidation of the function of noncoding disease-associated loci through neuroepigenomics is an important next step toward the development of testable hypotheses regarding biological processes involved in the pathogenesis of neuropsychiatric disorders.

Several lines of evidence suggest the involvement of a proportion of genome-wide associated variants in transcriptional regulatory mechanisms, including enrichment within expression quantitative trait loci (eQTL)[164–167] and modulation of *cis*-regulatory elements (CREs).[165,168–170] A CRE, such as a promoter, enhancer, or silencer, is a noncoding DNA sequence in, near, or distal to a gene that contains binding sites for regulatory factors and is required for proper spatiotemporal expression of the gene. The proposed mechanism is that disease-associated variants that lie within CREs affect the binding of regulatory proteins, such as transcription factors, leading to allele-specific differences in transcription and subsequent disease-related alterations in molecular pathways. The importance of neuroepigenomic annotations for informing the genetic influence on transcriptional regulation is supported by recent studies in human lymphoblastoid cell lines, where more than 50% of eQTLs were found to lie within CREs.[166]

Integration of neuroepigenomic annotations identifies a reduced set of functional SNPs to be tested for association with the disease in the GWAS datasets (Figs. 8.1 and 8.2). While still a young field with methods still under development, this approach bears significant promise because it would decrease the number of association tests from a couple of million variants (using imputed genotypes) to a couple of thousands. This approach leads to enhanced power by eliminating the excessive multiple testing corrections, (ii) identifying the true causal SNP out of nearby tag SNPs that are in linkage disequilibrium, and (iii) providing a plausible mechanistic and testable explanation for the effect of SNPs that can affect related functions, such as changes in gene expression, through allele-specific alterations in transcription factor binding site or alterations in the three-dimensional genome architecture associated with chromosomal loopings and transcriptional regulation in the brain.

4. SYNOPSIS AND OUTLOOK

Over the course of only a few years, we have witnessed a proliferation of epigenetic studies in the human brain, ranging from exploration of chromatin structures at a specific genomic locus to genome-wide epigenome mapping in defined cell types, generally with signal-to-noise ratios and signal quality comparable with those obtained in animal brains. Work from multiple groups, focusing mainly on human association cortex, points to large-scale remodeling of DNA and histone methylation landscapes during the late prenatal phase and early postnatal phase and early childhood,

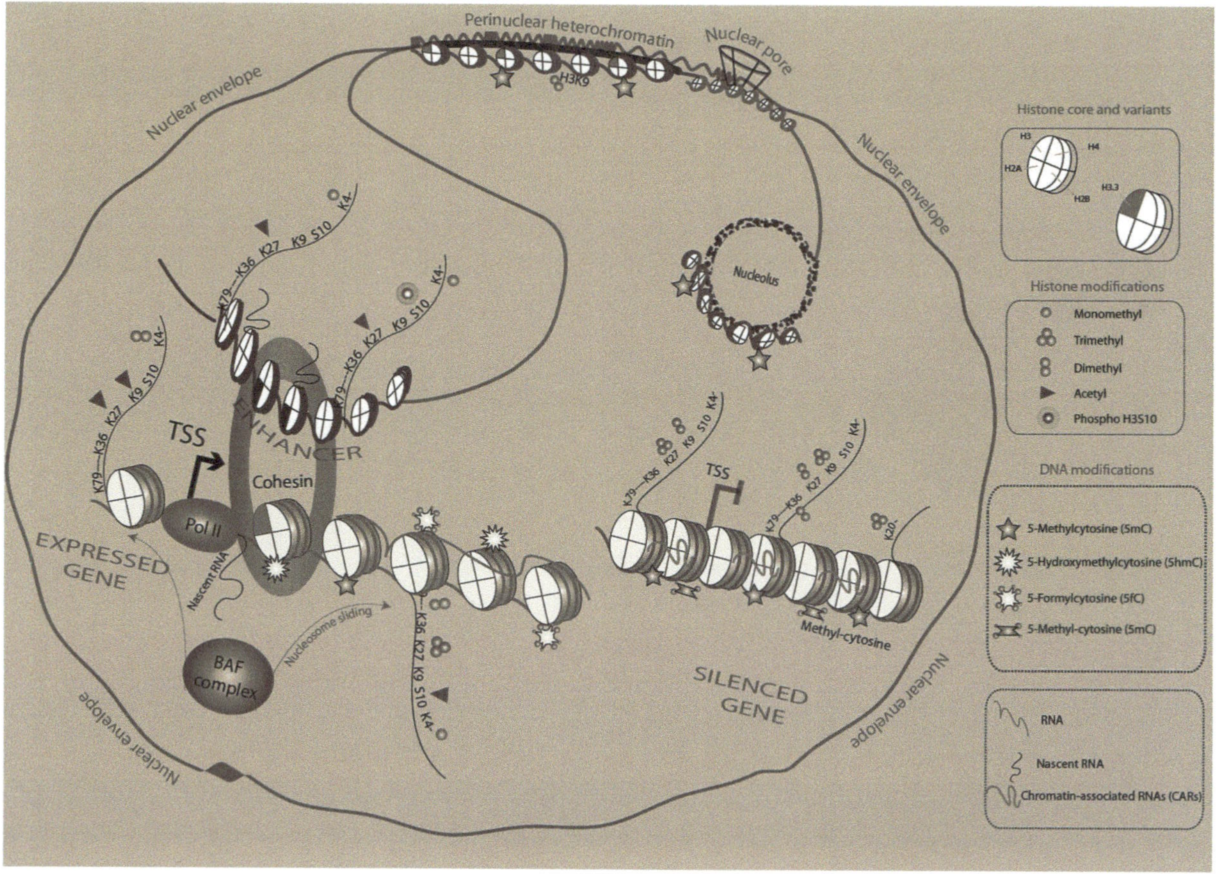

Figure 8.1 (See figure legend on next page)

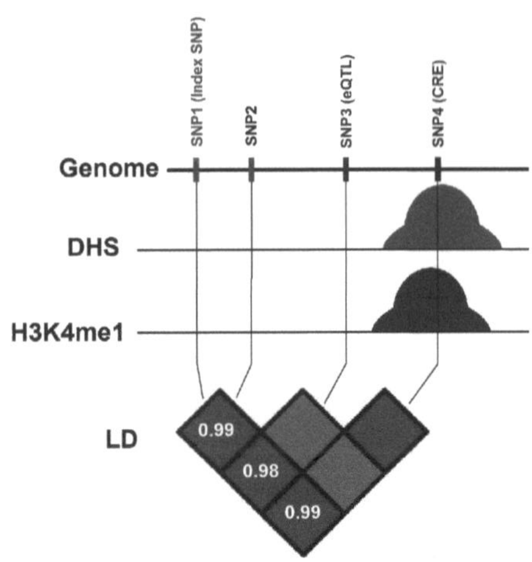

Figure 8.2 Neuroepigenomic annotations could prioritize variants with additional functional support. Example of four SNPs that are in strong LD ($r2 \approx 1$). SNP1 is the associated variant with a phenotype in a GWAS. SNP2 is an eQTL associated with changes in gene expression in a different study. SNP4 overlaps a DNase I hypersensitivity peak (a measure for open chromatin and accessibility of the DNA by regulatory proteins)[171,172] and a ChIP-seq peak for predicted enhancers (H3K4me1).[173,174] Therefore, multiple sources of evidence support that SNP4 is in a regulatory region that regulates gene expression of specific transcript and is associated with the phenotype of interest. (See the color plate.)

with comparatively less dramatic changes during subsequent stages of development and aging.[51,80,103] Still, hundreds of promoters are subject to epigenetic changes that seemingly continue into old age, and these data, taken together, leave little doubt that chromatin structures undergo remodeling throughout the life span of the human brain,[48,103,175] including neurons

Figure 8.1 The epigenome, from nucleus to nucleosome. Schematic illustration of (green) gene poised for transcription by polymerase II (Pol II) initiation complex, with nucleosome-free interval at transcription start site (TSS); (blue) distal enhancer sequence that in loop-like structure moves in close proximity to active gene; and (red) marks a small subset of heterochromatic portions of the genome, including silenced gene and heterochromatic structures bordering the nuclear envelope and pore complex and also the nucleolar periphery. A small subset of representative histone variants and H3 site-specific lysine (K) residues at N-terminal tail (K4, K9, K27, and K36) or core fold domain of the (histone) octamer (K79) and the H4K20 residue are shown as indicated, together with panel of mono- and trimethyl or acetyl modifications that differentiate between active promoters, transcribed gene bodies, and repressive chromatin, as indicated. DNA cytosines that are hydroxymethylated at the C5 position are mostly found at active promoters, while methylated cytosines are positioned within the body of actively transcribed genes and around repressed promoters and in constitutive heterochromatin. (See the color plate.)

and other terminally differentiated cells.[51] Based on postmortem brain work, epigenetic risk architectures are beginning to emerge for a number of common psychiatric conditions and disorders, including autism,[176] schizophrenia,[177] depression and bipolar disorder,[105,178] and alcoholism.[179] We predict that only very few, if any, loci will show replicated group-based differences when assayed in genome-wide epigenetic screens. Instead, we argue that epigenetic exploration of brain cells and tissue is ideally done in large cohorts, ideally hundreds, of phenotypically well-characterized subjects, in conjunction with next-generation sequencing-based epigenome and transcriptome profiling and high-coverage whole-genome sequencing of the same cases and tissues. Such type of datasets, when combined with the rapidly refining genetic risk maps for common neuropsychiatric diseases,[159,180,181] will provide a powerful tool to gain deep and unprecedented insights into the genomic foundations of cognition and emotion, including disorders unique to human.

CONFLICT OF INTEREST

The authors declare no conflict of interest.

ACKNOWLEDGMENT

Work conducted in the authors' laboratories is sponsored by the National Institutes of Health, the Veterans Administration, Autism Speaks, and the Brain & Behavior Research Foundation.

REFERENCES

1. Rodriguez-Paredes M, Esteller M. Cancer epigenetics reaches mainstream oncology. *Nat Med.* 2011;17(3):330–339.
2. Li G, Reinberg D. Chromatin higher-order structures and gene regulation. *Curr Opin Genet Dev.* 2011;21(2):175–186.
3. Kriaucionis S, Heintz N. The nuclear DNA base 5-hydroxymethylcytosine is present in Purkinje neurons and the brain. *Science.* 2009;324(5929):929–930.
4. Song CX, He C. Potential functional roles of DNA demethylation intermediates. *Trends Biochem Sci.* 2013;38(10):480–484.
5. Song CX, Yi C, He C. Mapping recently identified nucleotide variants in the genome and transcriptome. *Nat Biotechnol.* 2012;30(11):1107–1116.
6. Xie W, Barr CL, Kim A, et al. Base-resolution analyses of sequence and parent-of-origin dependent DNA methylation in the mouse genome. *Cell.* 2012;148(4):816–831.
7. Jin SG, Wu X, Li AX, Pfeifer GP. Genomic mapping of 5-hydroxymethylcytosine in the human brain. *Nucleic Acids Res.* 2011;39(12):5015–5024.
8. Song CX, Szulwach KE, Fu Y, et al. Selective chemical labeling reveals the genome-wide distribution of 5-hydroxymethylcytosine. *Nat Biotechnol.* 2011;29(1):68–72.
9. Mellen M, Ayata P, Dewell S, Kriaucionis S, Heintz N. MeCP2 binds to 5hmC enriched within active genes and accessible chromatin in the nervous system. *Cell.* 2012;151(7):1417–1430.

10. Khare T, Pai S, Koncevicius K, et al. 5-hmC in the brain is abundant in synaptic genes and shows differences at the exon-intron boundary. *Nat Struct Mol Biol.* 2012;19(10): 1037–1043.
11. Maunakea AK, Nagarajan RP, Bilenky M, et al. Conserved role of intragenic DNA methylation in regulating alternative promoters. *Nature.* 2010;466(7303):253–257.
12. Tan M, Luo H, Lee S, et al. Identification of 67 histone marks and histone lysine crotonylation as a new type of histone modification. *Cell.* 2011;146(6):1016–1028.
13. Kouzarides T. Chromatin modifications and their function. *Cell.* 2007;128(4): 693–705.
14. Taverna SD, Li H, Ruthenburg AJ, Allis CD, Patel DJ. How chromatin-binding modules interpret histone modifications: lessons from professional pocket pickers. *Nat Struct Mol Biol.* 2007;14(11):1025–1040.
15. Zhou VW, Goren A, Bernstein BE. Charting histone modifications and the functional organization of mammalian genomes. *Nat Rev Genet.* 2011;12(1):7–18.
16. Berger SL. The complex language of chromatin regulation during transcription. *Nature.* 2007;447(7143):407–412.
17. Mosammaparast N, Shi Y. Reversal of histone methylation: biochemical and molecular mechanisms of histone demethylases. *Annu Rev Biochem.* 2010;79:155–179.
18. Justin N, De Marco V, Aasland R, Gamblin SJ. Reading, writing and editing methylated lysines on histone tails: new insights from recent structural studies. *Curr Opin Struct Biol.* 2010;20(6):730–738.
19. Jin C, Felsenfeld G. Nucleosome stability mediated by histone variants H3.3 and H2A. Z. *Genes Dev.* 2007;21(12):1519–1529.
20. Ronan JL, Wu W, Crabtree GR. From neural development to cognition: unexpected roles for chromatin. *Nat Rev Genet.* 2013;14(5):347–359.
21. Vogel-Ciernia A, Wood MA. Neuron-specific chromatin remodeling: a missing link in epigenetic mechanisms underlying synaptic plasticity, memory, and intellectual disability disorders. *Neuropharmacology.* 2014;80:18–27.
22. Vogel-Ciernia A, Matheos DP, Barrett RM, et al. The neuron-specific chromatin regulatory subunit BAF53b is necessary for synaptic plasticity and memory. *Nat Neurosci.* 2013;16(5):552–561.
23. Rodriguez-Campos A, Azorin F. RNA is an integral component of chromatin that contributes to its structural organization. *PLoS One.* 2007;2(11):e1182.
24. Brockdorff N. Noncoding RNA and polycomb recruitment. *RNA.* 2013;19(4):429–442.
25. Zhao J, Sun BK, Erwin JA, Song JJ, Lee JT. Polycomb proteins targeted by a short repeat RNA to the mouse X chromosome. *Science.* 2008;322(5902):750–756.
26. Le Meur E, Watrin F, Landers M, Sturny R, Lalande M, Muscatelli F. Dynamic developmental regulation of the large non-coding RNA associated with the mouse 7C imprinted chromosomal region. *Dev Biol.* 2005;286(2):587–600.
27. Leung KN, Chamberlain SJ, Lalande M, LaSalle JM. Neuronal chromatin dynamics of imprinting in development and disease. *J Cell Biochem.* 2011;112(2):365–373.
28. Xin Z, Allis CD, Wagstaff J. Parent-specific complementary patterns of histone H3 lysine 9 and H3 lysine 4 methylation at the Prader-Willi syndrome imprinting center. *Am J Hum Genet.* 2001;69(6):1389–1394.
29. Shulha HP, Crisci JL, Reshetov D, et al. Human-specific histone methylation signatures at transcription start sites in prefrontal neurons. *PLoS Biol.* 2012;10(11):e1001427.
30. Cremer T, Cremer C. Chromosome territories, nuclear architecture and gene regulation in mammalian cells. *Nat Rev Genet.* 2001;2(4):292–301.
31. Duan Z, Andronescu M, Schutz K, et al. A three-dimensional model of the yeast genome. *Nature.* 2010;465(7296):363–367.
32. Wood AJ, Severson AF, Meyer BJ. Condensin and cohesin complexity: the expanding repertoire of functions. *Nat Rev Genet.* 2010;11(6):391–404.

33. Gaszner M, Felsenfeld G. Insulators: exploiting transcriptional and epigenetic mechanisms. *Nat Rev Genet.* 2006;7(9):703–713.
34. Moss JF, Oliver C, Berg K, Kaur G, Jephcott L, Cornish K. Prevalence of autism spectrum phenomenology in Cornelia de Lange and Cri du Chat syndromes. *Am J Ment Retard.* 2008;113(4):278–291.
35. Deardorff MA, Bando M, Nakato R, et al. HDAC8 mutations in Cornelia de Lange syndrome affect the cohesin acetylation cycle. *Nature.* 2012;489(7415): 313–317.
36. Gervasini C, Parenti I, Picinelli C, et al. Molecular characterization of a mosaic NIPBL deletion in a Cornelia de Lange patient with severe phenotype. *Eur J Med Genet.* 2013;56(3):138–143.
37. Kagey MH, Newman JJ, Bilodeau S, et al. Mediator and cohesin connect gene expression and chromatin architecture. *Nature.* 2010;467(7314):430–435.
38. Huang HS, Matevossian A, Jiang Y, Akbarian S. Chromatin immunoprecipitation in postmortem brain. *J Neurosci Methods.* 2006;156(1–2):284–292.
39. Adli M, Bernstein BE. Whole-genome chromatin profiling from limited numbers of cells using nano-ChIP-seq. *Nat Protoc.* 2011;6(10):1656–1668.
40. Abdolmaleky HM, Cheng KH, Faraone SV, et al. Hypomethylation of MB-COMT promoter is a major risk factor for schizophrenia and bipolar disorder. *Hum Mol Genet.* 2006;15(21):3132–3145.
41. Abdolmaleky HM, Cheng KH, Russo A, et al. Hypermethylation of the reelin (RELN) promoter in the brain of schizophrenic patients: a preliminary report. *Am J Med Genet B Neuropsychiatr Genet.* 2005;134B(1):60–66.
42. Grayson DR, Jia X, Chen Y, et al. Reelin promoter hypermethylation in schizophrenia. *Proc Natl Acad Sci USA.* 2005;102(26):9341–9346.
43. Huang HS, Matevossian A, Whittle C, et al. Prefrontal dysfunction in schizophrenia involves mixed-lineage leukemia 1-regulated histone methylation at GABAergic gene promoters. *J Neurosci.* 2007;27(42):11254–11262.
44. Iwamoto K, Bundo M, Yamada K, et al. DNA methylation status of SOX10 correlates with its downregulation and oligodendrocyte dysfunction in schizophrenia. *J Neurosci.* 2005;25(22):5376–5381.
45. Mill J, Tang T, Kaminsky Z, et al. Epigenomic profiling reveals DNA-methylation changes associated with major psychosis. *Am J Hum Genet.* 2008;82(3):696–711.
46. Tochigi M, Iwamoto K, Bundo M, et al. Methylation status of the reelin promoter region in the brain of schizophrenic patients. *Biol Psychiatry.* 2008;63(5): 530–533.
47. Tamura Y, Kunugi H, Ohashi J, Hohjoh H. Epigenetic aberration of the human REELIN gene in psychiatric disorders. *Mol Psychiatry.* 2007;12(6):519, 593–600.
48. Siegmund KD, Connor CM, Campan M, et al. DNA methylation in the human cerebral cortex is dynamically regulated throughout the life span and involves differentiated neurons. *PLoS One.* 2007;2(9):e895.
49. Iwamoto K, Bundo M, Ueda J, et al. Neurons show distinctive DNA methylation profile and higher interindividual variations compared with non-neurons. *Genome Res.* 2011;21(5):688–696.
50. Sugawara H, Iwamoto K, Bundo M, Ueda J, Ishigooka J, Kato T. Comprehensive DNA methylation analysis of human peripheral blood leukocytes and lymphoblastoid cell lines. *Epigenetics.* 2011;6(4):508–515.
51. Cheung I, Shulha HP, Jiang Y, et al. Developmental regulation and individual differences of neuronal H3K4me3 epigenomes in the prefrontal cortex. *Proc Natl Acad Sci USA.* 2010;107(19):8824–8829.
52. Montano CM, Irizarry RA, Kaufmann WE, et al. Measuring cell-type specific differential methylation in human brain tissue. *Genome Biol.* 2013;14(8):R94.

53. Guintivano J, Aryee MJ, Kaminsky ZA. A cell epigenotype specific model for the correction of brain cellular heterogeneity bias and its application to age, brain region and major depression. *Epigenetics*. 2013;8(3):290–302.
54. Matevossian A, Akbarian S. Neuronal nuclei isolation from human postmortem brain tissue. *J Vis Exp*. 2008;20:914.
55. Sim FJ, McClain CR, Schanz SJ, Protack TL, Windrem MS, Goldman SA. CD140a identifies a population of highly myelinogenic, migration-competent and efficiently engrafting human oligodendrocyte progenitor cells. *Nat Biotechnol*. 2011;29(10):934–941.
56. Sundberg M, Skottman H, Suuronen R, Narkilahti S. Production and isolation of NG2+ oligodendrocyte precursors from human embryonic stem cells in defined serum-free medium. *Stem Cell Res*. 2010;5(2):91–103.
57. Nunes MC, Roy NS, Keyoung HM, et al. Identification and isolation of multipotential neural progenitor cells from the subcortical white matter of the adult human brain. *Nat Med*. 2003;9(4):439–447.
58. Jungblut M, Tiveron MC, Barral S, et al. Isolation and characterization of living primary astroglial cells using the new GLAST-specific monoclonal antibody ACSA-1. *Glia*. 2012;60(6):894–907.
59. Yuan SH, Martin J, Elia J, et al. Cell-surface marker signatures for the isolation of neural stem cells, glia and neurons derived from human pluripotent stem cells. *PLoS One*. 2011;6(3):e17540.
60. Uchida N, Buck DW, He D, et al. Direct isolation of human central nervous system stem cells. *Proc Natl Acad Sci USA*. 2000;97(26):14720–14725.
61. Melief J, Koning N, Schuurman KG, et al. Phenotyping primary human microglia: tight regulation of LPS responsiveness. *Glia*. 2012;60(10):1506–1517.
62. Larsson HM, Lee ST, Roccio M, et al. Sorting live stem cells based on Sox2 mRNA expression. *PLoS One*. 2012;7(11):e49874.
63. Codega P, Silva-Vargas V, Paul A, et al. Prospective identification and purification of quiescent adult neural stem cells from their in vivo niche. *Neuron*. 2014;82(3):545–559.
64. Matevossian A, Akbarian S. A chromatin assay for human brain tissue. *J Vis Exp*. 2008;13:717.
65. Jiang Y, Matevossian A, Huang HS, Straubhaar J, Akbarian S. Isolation of neuronal chromatin from brain tissue. *BMC Neurosci*. 2008;9:42.
66. Hayashi Y, Nihonmatsu-Kikuchi N, Yu X, Ishimoto K, Hisanaga SI, Tatebayashi Y. A novel, rapid, quantitative cell-counting method reveals oligodendroglial reduction in the frontopolar cortex in major depressive disorder. *Mol Psychiatry*. 2011;16(12):1155–1158.
67. Ernst C, McGowan PO, Deleva V, Meaney MJ, Szyf M, Turecki G. The effects of pH on DNA methylation state: in vitro and post-mortem brain studies. *J Neurosci Methods*. 2008;174(1):123–125.
68. Mitchell AC, Bharadwaj R, Whittle C, et al. The genome in three dimensions: a new frontier in human brain research. *Biol Psychiatry*. 2013;75(12):961–969.
69. Monoranu CM, Grunblatt E, Bartl J, et al. Methyl- and acetyltransferases are stable epigenetic markers postmortem. *Cell Tissue Bank*. 2011;12(4):289–297.
70. Rogaev EI, Lukiw WJ, Lavrushina O, Rogaeva EA, St George-Hyslop PH. The upstream promoter of the beta-amyloid precursor protein gene (APP) shows differential patterns of methylation in human brain. *Genomics*. 1994;22(2):340–347.
71. West RL, Lee JM, Maroun LE. Hypomethylation of the amyloid precursor protein gene in the brain of an Alzheimer's disease patient. *J Mol Neurosci*. 1995;6(2):141–146.
72. Tohgi H, Utsugisawa K, Nagane Y, Yoshimura M, Genda Y, Ukitsu M. Reduction with age in methylcytosine in the promoter region −224 approximately −101 of the amyloid precursor protein gene in autopsy human cortex. *Brain Res Mol Brain Res*. 1999;70(2):288–292.

73. Tassone F, Hagerman RJ, Gane LW, Taylor AK. Strong similarities of the FMR1 mutation in multiple tissues: postmortem studies of a male with a full mutation and a male carrier of a premutation. *Am J Med Genet*. 1999;84(3):240–244.
74. Stadler F, Kolb G, Rubusch L, Baker SP, Jones EG, Akbarian S. Histone methylation at gene promoters is associated with developmental regulation and region-specific expression of ionotropic and metabotropic glutamate receptors in human brain. *J Neurochem*. 2005;94(2):324–336.
75. Oberle I, Rousseau F, Heitz D, et al. Instability of a 550-base pair DNA segment and abnormal methylation in fragile X syndrome. *Science*. 1991;252(5009):1097–1102.
76. Zhao J, Grant SF. Advances in whole genome sequencing technology. *Curr Pharm Biotechnol*. 2011;12(2):293–305.
77. Park PJ. ChIP-seq: advantages and challenges of a maturing technology. *Nat Rev Genet*. 2009;10(10):669–680.
78. Thurman RE, Rynes E, Humbert R, et al. The accessible chromatin landscape of the human genome. *Nature*. 2012;489(7414):75–82.
79. Bernstein BE, Birney E, Dunham I, Green ED, Gunter C, Snyder M. An integrated encyclopedia of DNA elements in the human genome. *Nature*. 2012;489(7414):57–74.
80. Lister R, Mukamel EA, Nery JR, et al. Global epigenomic reconfiguration during mammalian brain development. *Science*. 2013;341(6146):1237905.
81. Zhu J, Adli M, Zou JY, et al. Genome-wide chromatin state transitions associated with developmental and environmental cues. *Cell*. 2013;152(3):642–654.
82. Shulha HP, Cheung I, Guo Y, Akbarian S, Weng Z. Coordinated cell type-specific epigenetic remodeling in prefrontal cortex begins before birth and continues into early adulthood. *PLoS Genet*. 2013;9(4):e1003433.
83. Mitchell AC, Bharadwaj R, Whittle C, et al. The genome in three dimensions: a new frontier in human brain research. *Biol Psychiatry*. 2014;75(12):961–969.
84. Bharadwaj R, Jiang Y, Mao W, et al. Conserved chromosome 2q31 conformations are associated with transcriptional regulation of GAD1 GABA synthesis enzyme and altered in prefrontal cortex of subjects with schizophrenia. *J Neurosci*. 2013;33(29): 11839–11851.
85. Dekker J, Marti-Renom MA, Mirny LA. Exploring the three-dimensional organization of genomes: interpreting chromatin interaction data. *Nat Rev Genet*. 2013;14(6): 390–403.
86. van Berkum NL, Dekker J. Determining spatial chromatin organization of large genomic regions using 5C technology. *Methods Mol Biol*. 2009;567:189–213.
87. Simonis M, Klous P, Splinter E, et al. Nuclear organization of active and inactive chromatin domains uncovered by chromosome conformation capture-on-chip (4C). *Nat Genet*. 2006;38(11):1348–1354.
88. Dostie J, Richmond TA, Arnaout RA, et al. Chromosome conformation capture carbon copy (5C): a massively parallel solution for mapping interactions between genomic elements. *Genome Res*. 2006;16(10):1299–1309.
89. Lieberman-Aiden E, van Berkum NL, Williams L, et al. Comprehensive mapping of long-range interactions reveals folding principles of the human genome. *Science*. 2009;326(5950):289–293.
90. Li G, Fullwood MJ, Xu H, et al. ChIA-PET tool for comprehensive chromatin interaction analysis with paired-end tag sequencing. *Genome Biol*. 2010;11(2):R22.
91. Gehlen LR, Gruenert G, Jones MB, Rodley CD, Langowski J, O'Sullivan JM. Chromosome positioning and the clustering of functionally related loci in yeast is driven by chromosomal interactions. *Nucleus*. 2012;3(4):370–383.
92. Cagliero C, Grand RS, Jones MB, Jin DJ, O'Sullivan JM. Genome conformation capture reveals that the Escherichia coli chromosome is organized by replication and transcription. *Nucleic Acids Res*. 2013;41(12):6058–6071.

93. Tanizawa H, Iwasaki O, Tanaka A, et al. Mapping of long-range associations throughout the fission yeast genome reveals global genome organization linked to transcriptional regulation. *Nucleic Acids Res*. 2010;38(22):8164–8177.
94. Shen Y, Yue F, McCleary DF, et al. A map of the cis-regulatory sequences in the mouse genome. *Nature*. 2012;488(7409):116–120.
95. Jin F, Li Y, Dixon JR, et al. A high-resolution map of the three-dimensional chromatin interactome in human cells. *Nature*. 2013;503(7475):290–294.
96. Gheldof N, Leleu M, Noordermeer D, Rougemont J, Reymond A. Detecting long-range chromatin interactions using the chromosome conformation capture sequencing (4C-seq) method. *Methods Mol Biol*. 2012;786:211–225.
97. Stadhouders R, Kolovos P, Brouwer R, et al. Multiplexed chromosome conformation capture sequencing for rapid genome-scale high-resolution detection of long-range chromatin interactions. *Nat Protoc*. 2013;8(3):509–524.
98. Zhang TY, Hellstrom IC, Bagot RC, Wen X, Diorio J, Meaney MJ. Maternal care and DNA methylation of a glutamic acid decarboxylase 1 promoter in rat hippocampus. *J Neurosci*. 2010;30(39):13130–13137.
99. McGowan PO, Sasaki A, D'Alessio AC, et al. Epigenetic regulation of the glucocorticoid receptor in human brain associates with childhood abuse. *Nat Neurosci*. 2009;12(3):342–348.
100. McGowan PO, Sasaki A, Huang TC, et al. Promoter-wide hypermethylation of the ribosomal RNA gene promoter in the suicide brain. *PLoS One*. 2008;3(5):e2085.
101. Erraji-Benchekroun L, Underwood MD, Arango V, et al. Molecular aging in human prefrontal cortex is selective and continuous throughout adult life. *Biol Psychiatry*. 2005;57(5):549–558.
102. Tang B, Chang WL, Lanigan CM, Dean B, Sutcliffe JG, Thomas EA. Normal human aging and early-stage schizophrenia share common molecular profiles. *Aging Cell*. 2009;8(3):339–342.
103. Numata S, Ye T, Hyde TM, et al. DNA methylation signatures in development and aging of the human prefrontal cortex. *Am J Hum Genet*. 2012;90(2):260–272.
104. Al-Mahdawi S, Pinto RM, Ismail O, et al. The Friedreich ataxia GAA repeat expansion mutation induces comparable epigenetic changes in human and transgenic mouse brain and heart tissues. *Hum Mol Genet*. 2008;17(5):735–746.
105. Gamazon ER, Badner JA, Cheng L, et al. Enrichment of cis-regulatory gene expression SNPs and methylation quantitative trait loci among bipolar disorder susceptibility variants. *Mol Psychiatry*. 2012;18(3):340–346.
106. Kaminsky ZA, Tang T, Wang SC, et al. DNA methylation profiles in monozygotic and dizygotic twins. *Nat Genet*. 2009;41(2):240–245.
107. Kundakovic M, Chen Y, Costa E, Grayson DR. DNA methyltransferase inhibitors coordinately induce expression of the human reelin and glutamic acid decarboxylase 67 genes. *Mol Pharmacol*. 2007;71(3):644–653.
108. Levenson JM, Roth TL, Lubin FD, et al. Evidence that DNA (cytosine-5) methyltransferase regulates synaptic plasticity in the hippocampus. *J Biol Chem*. 2006;281(23):15763–15773.
109. Martinowich K, Hattori D, Wu H, et al. DNA methylation-related chromatin remodeling in activity-dependent BDNF gene regulation. *Science*. 2003;302(5646): 890–893.
110. Nelson ED, Kavalali ET, Monteggia LM. Activity-dependent suppression of miniature neurotransmission through the regulation of DNA methylation. *J Neurosci*. 2008;28(2):395–406.
111. Guo JU, Ma DK, Mo H, et al. Neuronal activity modifies the DNA methylation landscape in the adult brain. *Nat Neurosci*. 2011;14(10):1345–1351.
112. Ma DK, Jang MH, Guo JU, et al. Neuronal activity-induced Gadd45b promotes epigenetic DNA demethylation and adult neurogenesis. *Science*. 2009;323(5917):1074–1077.

113. Cohen S, Gabel HW, Hemberg M, et al. Genome-wide activity-dependent MeCP2 phosphorylation regulates nervous system development and function. *Neuron.* 2011;72(1):72–85.
114. Li H, Zhong X, Chau KF, Williams EC, Chang Q. Loss of activity-induced phosphorylation of MeCP2 enhances synaptogenesis, LTP and spatial memory. *Nat Neurosci.* 2011;14(8):1001–1008.
115. Day JJ, Sweatt JD. DNA methylation and memory formation. *Nat Neurosci.* 2010;13(11):1319–1323.
116. Miller CA, Gavin CF, White JA, et al. Cortical DNA methylation maintains remote memory. *Nat Neurosci.* 2010;13(6):664–666.
117. Ooi SK, Bestor TH. The colorful history of active DNA demethylation. *Cell.* 2008;133(7):1145–1148.
118. Klose RJ, Zhang Y. Regulation of histone methylation by demethylimination and demethylation. *Nat Rev Mol Cell Biol.* 2007;8(4):307–318.
119. Loenarz C, Schofield CJ. Physiological and biochemical aspects of hydroxylations and demethylations catalyzed by human 2-oxoglutarate oxygenases. *Trends Biochem Sci.* 2011;36(1):7–18.
120. Miller CA, Sweatt JD. Covalent modification of DNA regulates memory formation. *Neuron.* 2007;53(6):857–869.
121. Endres M, Meisel A, Biniszkiewicz D, et al. DNA methyltransferase contributes to delayed ischemic brain injury. *J Neurosci.* 2000;20(9):3175–3181.
122. Bollati V, Baccarelli A, Hou L, et al. Changes in DNA methylation patterns in subjects exposed to low-dose benzene. *Cancer Res.* 2007;67(3):876–880.
123. Desaulniers D, Xiao GH, Leingartner K, Chu I, Musicki B, Tsang BK. Comparisons of brain, uterus, and liver mRNA expression for cytochrome p450s, DNA methyltransferase-1, and catechol-o-methyltransferase in prepubertal female Sprague–Dawley rats exposed to a mixture of aryl hydrocarbon receptor agonists. *Toxicol Sci.* 2005;86(1):175–184.
124. Salnikow K, Zhitkovich A. Genetic and epigenetic mechanisms in metal carcinogenesis and cocarcinogenesis: nickel, arsenic, and chromium. *Chem Res Toxicol.* 2008;21(1):28–44.
125. Satta R, Maloku E, Zhubi A, et al. Nicotine decreases DNA methyltransferase 1 expression and glutamic acid decarboxylase 67 promoter methylation in GABAergic interneurons. *Proc Natl Acad Sci USA.* 2008;105(42):16356–16361.
126. Satta R, Maloku E, Costa E, Guidotti A. Stimulation of brain nicotinic acetylcholine receptors (nAChRs) decreases DNA methyltransferase 1 (DNMT1) expression in cortical and hippocampal GABAergic neurons of Swiss albino mice. *Soc Neurosci Abstr.* 2007.
127. Marutha Ravindran CR, Ticku MK. Changes in methylation pattern of NMDA receptor NR2B gene in cortical neurons after chronic ethanol treatment in mice. *Brain Res Mol Brain Res.* 2004;121(1–2):19–27.
128. Numachi Y, Shen H, Yoshida S, et al. Methamphetamine alters expression of DNA methyltransferase 1 mRNA in rat brain. *Neurosci Lett.* 2007;414(3):213–217.
129. Numachi Y, Yoshida S, Yamashita M, et al. Psychostimulant alters expression of DNA methyltransferase mRNA in the rat brain. *Ann N Y Acad Sci.* 2004;1025:102–109.
130. LaPlant Q, Vialou V, Covington 3rd HE, et al. Dnmt3a regulates emotional behavior and spine plasticity in the nucleus accumbens. *Nat Neurosci.* 2010;13(9):1137–1143.
131. Cheng MC, Liao DL, Hsiung CA, Chen CY, Liao YC, Chen CH. Chronic treatment with aripiprazole induces differential gene expression in the rat frontal cortex. *Int J Neuropsychopharmacol.* 2008;11(2):207–216.
132. Shimabukuro M, Jinno Y, Fuke C, Okazaki Y. Haloperidol treatment induces tissue- and sex-specific changes in DNA methylation: a control study using rats. *Behav Brain Funct.* 2006;2:37.

133. Li J, Guo Y, Schroeder FA, et al. Dopamine D2-like antagonists induce chromatin remodeling in striatal neurons through cyclic AMP-protein kinase A and NMDA receptor signaling. *J Neurochem*. 2004;90(5):1117–1131.
134. Dong E, Nelson M, Grayson DR, Costa E, Guidotti A. Clozapine and sulpiride but not haloperidol or olanzapine activate brain DNA demethylation. *Proc Natl Acad Sci USA*. 2008;105(36):13614–13619.
135. Kwon B, Houpt TA. Phospho-acetylation of histone H3 in the amygdala after acute lithium chloride. *Brain Res*. 2010;1333:36–47.
136. Bredy TW, Wu H, Crego C, Zellhoefer J, Sun YE, Barad M. Histone modifications around individual BDNF gene promoters in prefrontal cortex are associated with extinction of conditioned fear. *Learn Mem*. 2007;14(4):268–276.
137. Takahashi K, Yamanaka S. Induction of pluripotent stem cells from mouse embryonic and adult fibroblast cultures by defined factors. *Cell*. 2006;126(4):663–676.
138. Kim KS. Induced pluripotent stem (iPS) cells and their future in psychiatry. *Neuropsychopharmacology*. 2010;35(1):346–348.
139. Lancaster MA, Renner M, Martin CA, et al. Cerebral organoids model human brain development and microcephaly. *Nature*. 2013;501(7467):373–379.
140. Shcheglovitov A, Shcheglovitova O, Yazawa M, et al. SHANK3 and IGF1 restore synaptic deficits in neurons from 22q13 deletion syndrome patients. *Nature*. 2013;503(7475):267–271.
141. Farra N, Zhang WB, Pasceri P, Eubanks JH, Salter MW, Ellis J. Rett syndrome induced pluripotent stem cell-derived neurons reveal novel neurophysiological alterations. *Mol Psychiatry*. 2012;17(12):1261–1271.
142. Brennand KJ, Simone A, Jou J, et al. Modelling schizophrenia using human induced pluripotent stem cells. *Nature*. 2011;473(7346):221–225.
143. Pang ZP, Yang N, Vierbuchen T, et al. Induction of human neuronal cells by defined transcription factors. *Nature*. 2011;476(7359):220–223.
144. Stevens SJ, van Ravenswaaij-Arts CM, Janssen JW, et al. MYT1L is a candidate gene for intellectual disability in patients with 2p25.3 (2pter) deletions. *Am J Med Genet A*. 2011;155A(11):2739–2745.
145. Muotri AR, Marchetto MC, Coufal NG, et al. L1 retrotransposition in neurons is modulated by MeCP2. *Nature*. 2010;468(7322):443–446.
146. Skene PJ, Illingworth RS, Webb S, et al. Neuronal MeCP2 is expressed at near histone-octamer levels and globally alters the chromatin state. *Mol Cell*. 2010;37(4):457–468.
147. Amir RE, Van den Veyver IB, Wan M, Tran CQ, Francke U, Zoghbi HY. Rett syndrome is caused by mutations in X-linked MECP2, encoding methyl-CpG-binding protein 2. *Nat Genet*. 1999;23(2):185–188.
148. Chouery E, Ghoch JA, Corbani S, et al. A novel deletion in ZBTB24 in a Lebanese family with Immunodeficiency, centromeric instability, and facial anomalies syndrome type 2. *Clin Genet*. 2012;82(5):489–493.
149. Huangfu D, Osafune K, Maehr R, et al. Induction of pluripotent stem cells from primary human fibroblasts with only Oct4 and Sox2. *Nat Biotechnol*. 2008;26(11):1269–1275.
150. Yu IT, Park JY, Kim SH, Lee JS, Kim YS, Son H. Valproic acid promotes neuronal differentiation by induction of proneural factors in association with H4 acetylation. *Neuropharmacology*. 2009;56(2):473–480.
151. Hsieh J, Nakashima K, Kuwabara T, Mejia E, Gage FH. Histone deacetylase inhibition-mediated neuronal differentiation of multipotent adult neural progenitor cells. *Proc Natl Acad Sci USA*. 2004;101(47):16659–16664.
152. Guidotti A, Dong E, Kundakovic M, Satta R, Grayson DR, Costa E. Characterization of the action of antipsychotic subtypes on valproate-induced chromatin remodeling. *Trends Pharmacol Sci*. 2009;30(2):55–60.

153. Gavin DP, Kartan S, Chase K, Jayaraman S, Sharma RP. Histone deacetylase inhibitors and candidate gene expression: an in vivo and in vitro approach to studying chromatin remodeling in a clinical population. *J Psychiatr Res*. 2009;43(9):870–876.
154. Ohi Y, Qin H, Hong C, et al. Incomplete DNA methylation underlies a transcriptional memory of somatic cells in human iPS cells. *Nat Cell Biol*. 2011;13(5):541–549.
155. Nishino K, Toyoda M, Yamazaki-Inoue M, et al. DNA methylation dynamics in human induced pluripotent stem cells over time. *PLoS Genet*. 2011;7(5):e1002085.
156. Tobin SC, Kim K. Generating pluripotent stem cells: differential epigenetic changes during cellular reprogramming. *FEBS Lett*. 2012;586(18):2874–2881.
157. Tobe BT, Brandel MG, Nye JS, Snyder EY. Implications and limitations of cellular reprogramming for psychiatric drug development. *Exp Mol Med*. 2013;45:e59.
158. Gavin DP, Kartan S, Chase K, Grayson DR, Sharma RP. Reduced baseline acetylated histone 3 levels, and a blunted response to HDAC inhibition in lymphocyte cultures from schizophrenia subjects. *Schizophr Res*. 2008;103(1–3):330–332.
159. Ripke S, O'Dushlaine C, Chambert K, et al. Genome-wide association analysis identifies 13 new risk loci for schizophrenia. *Nat Genet*. 2013;45(10):1150–1159.
160. Sklar P, Ripke S, Scott LJ, et al. Large-scale genome-wide association analysis of bipolar disorder identifies a new susceptibility locus near ODZ4. *Nat Genet*. 2011;43(10):977–983.
161. Purcell SM, Wray NR, Stone JL, et al. Common polygenic variation contributes to risk of schizophrenia and bipolar disorder. *Nature*. 2009;460(7256):748–752.
162. Shi J, Levinson DF, Duan J, et al. Common variants on chromosome 6p22.1 are associated with schizophrenia. *Nature*. 2009;460(7256):753–757.
163. Stefansson H, Ophoff RA, Steinberg S, et al. Common variants conferring risk of schizophrenia. *Nature*. 2009;460(7256):744–747.
164. Zhong H, Beaulaurier J, Lum PY, et al. Liver and adipose expression associated SNPs are enriched for association to type 2 diabetes. *PLoS Genet*. 2010;6:e1000932.
165. Cookson W, Liang L, Abecasis G, Moffatt M, Lathrop M. Mapping complex disease traits with global gene expression. *Nat Rev Genet*. 2009;10(3):184–194.
166. Degner JF, Pai AA, Pique-Regi R, et al. DNase I sensitivity QTLs are a major determinant of human expression variation. *Nature*. 2012;482(7385):390–394.
167. Nicolae DL, Gamazon E, Zhang W, Duan S, Dolan ME, Cox NJ. Trait-associated SNPs are more likely to be eQTLs: annotation to enhance discovery from GWAS. *PLoS Genet*. 2010;6(4):e1000888.
168. Musunuru K, Strong A, Frank-Kamenetsky M, et al. From noncoding variant to phenotype via SORT1 at the 1p13 cholesterol locus. *Nature*. 2010;466(7307):714–719.
169. Harismendy O, Notani D, Song X, et al. 9p21 DNA variants associated with coronary artery disease impair interferon-gamma signalling response. *Nature*. 2011;470(7333):264–268.
170. Maurano MT, Humbert R, Rynes E, et al. Systematic localization of common disease-associated variation in regulatory DNA. *Science*. 2012;337(6099):1190–1195.
171. Karnik R, Meissner A. Browsing (Epi)genomes: a guide to data resources and epigenome browsers for stem cell researchers. *Cell Stem Cell*. 2013;13(1):14–21.
172. Ballare C, Zaurin R, Vicent GP, Beato M. More help than hindrance: nucleosomes aid transcriptional regulation. *Nucleus*. 2013;4(3):189–194.
173. Aday AW, Zhu LJ, Lakshmanan A, Wang J, Lawson ND. Identification of cis regulatory features in the embryonic zebrafish genome through large-scale profiling of H3K4me1 and H3K4me3 binding sites. *Dev Biol*. 2011;357(2):450–462.
174. Rada-Iglesias A, Bajpai R, Swigut T, Brugmann SA, Flynn RA, Wysocka J. A unique chromatin signature uncovers early developmental enhancers in humans. *Nature*. 2011;470(7333):279–283.

175. Hernandez DG, Nalls MA, Gibbs JR, et al. Distinct DNA methylation changes highly correlated with chronological age in the human brain. *Hum Mol Genet.* 2011;20(6): 1164–1172.
176. Shulha HP, Cheung I, Whittle C, et al. Epigenetic signatures of autism: trimethylated H3K4 landscapes in prefrontal neurons. *Arch Gen Psychiatry.* 2011;69(3):314–324.
177. Akbarian S. Epigenetics of schizophrenia. *Curr Top Behav Neurosci.* 2010;4:611–628.
178. Tang B, Dean B, Thomas EA. Disease- and age-related changes in histone acetylation at gene promoters in psychiatric disorders. *Transl Psychiatry.* 2011;1:e64.
179. Taqi MM, Bazov I, Watanabe H, et al. Prodynorphin CpG-SNPs associated with alcohol dependence: elevated methylation in the brain of human alcoholics. *Addict Biol.* 2011;16(3):499–509.
180. Purcell SM, Moran JL, Fromer M, et al. A polygenic burden of rare disruptive mutations in schizophrenia. *Nature.* 2014;506(7487):185–190.
181. Fromer M, Pocklington AJ, Kavanagh DH, et al. De novo mutations in schizophrenia implicate synaptic networks. *Nature.* 2014;506(7487):179–184.

INDEX

Note: Page numbers followed by "*f*" indicate figures and "*t*" indicate tables.

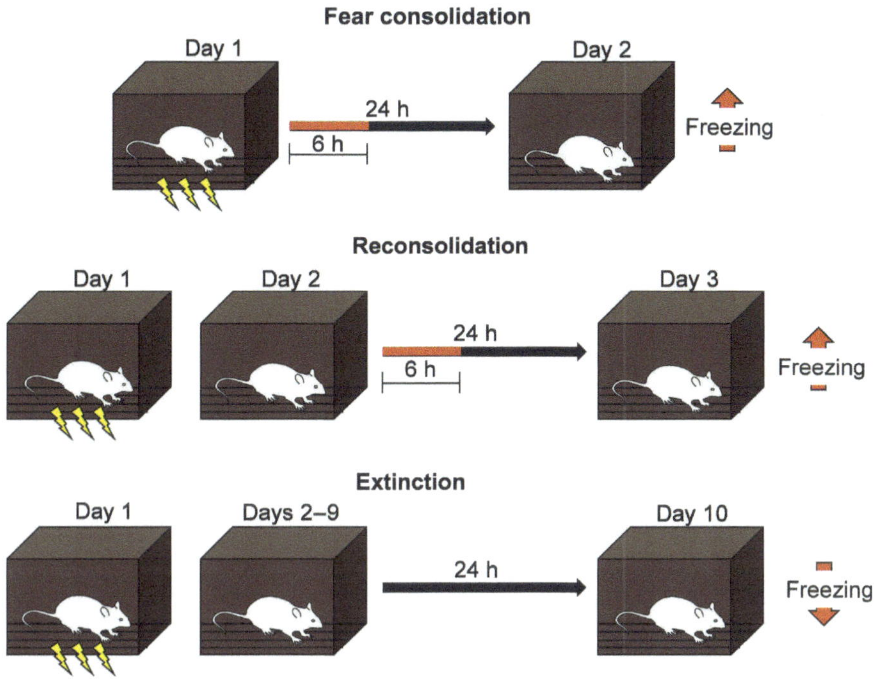

Timothy J. Jarome *et al.*, Figure 1.1

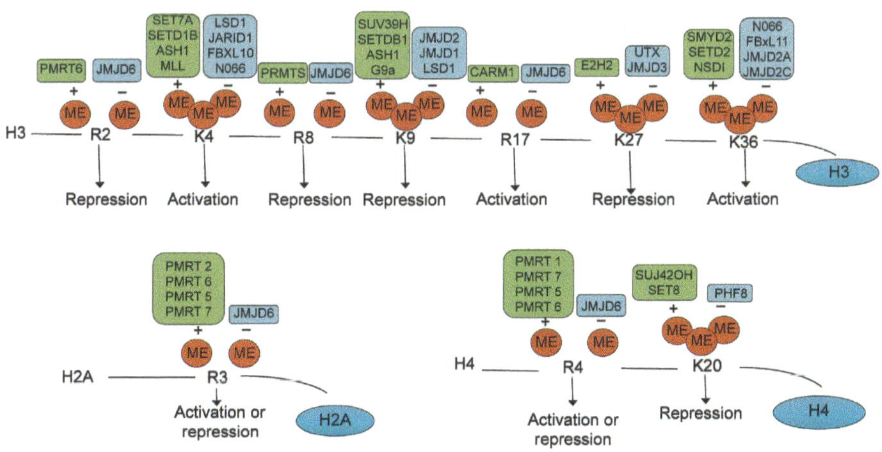

Timothy J. Jarome *et al.*, Figure 1.2

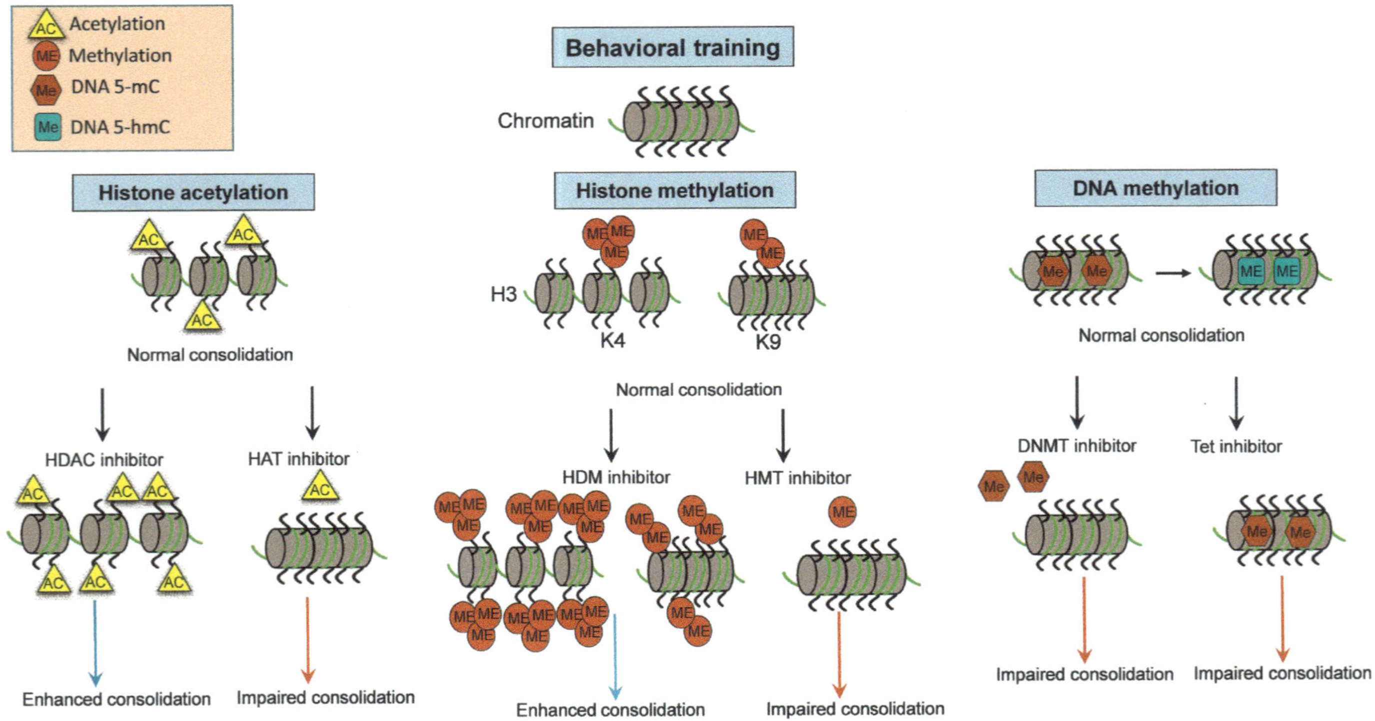

TIMOTHY J. JAROME *ET AL.*, FIGURE 1.3

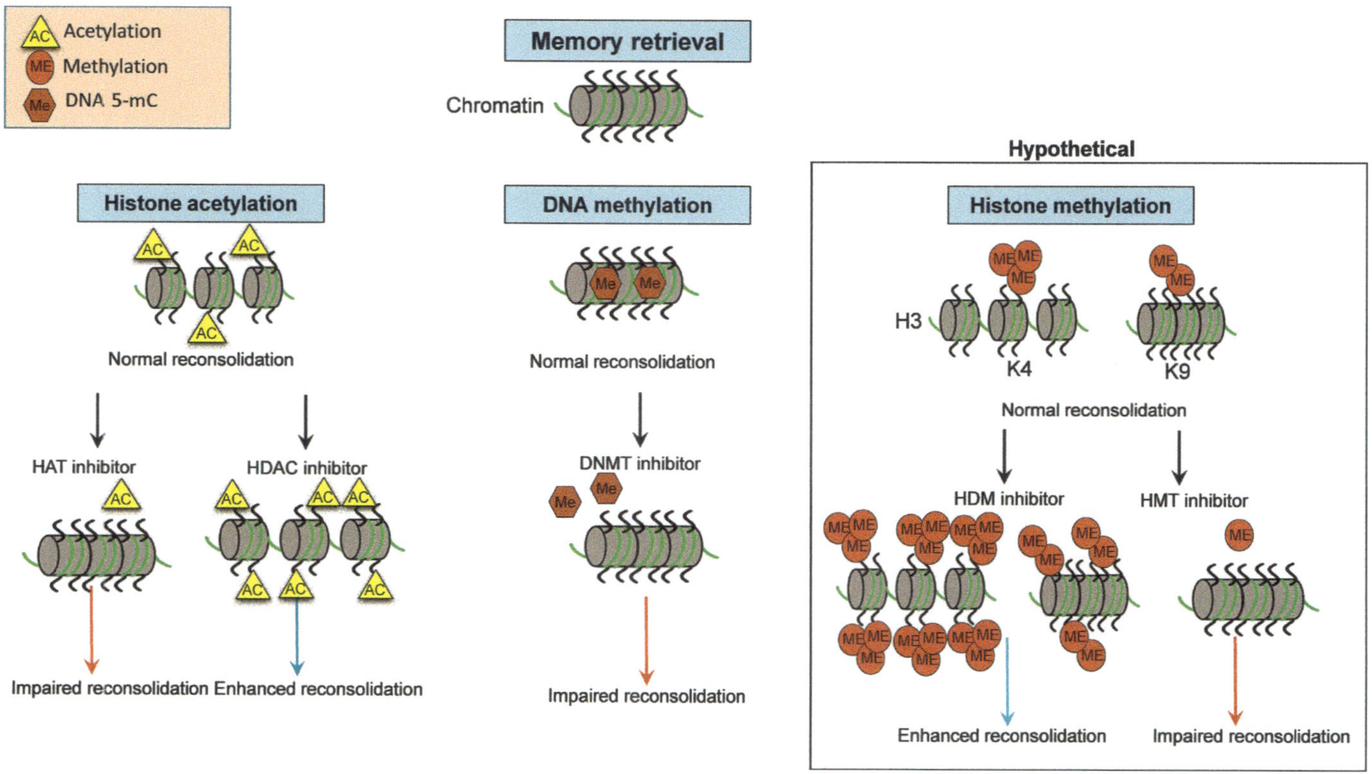

TIMOTHY J. JAROME *ET AL.*, FIGURE 1.4

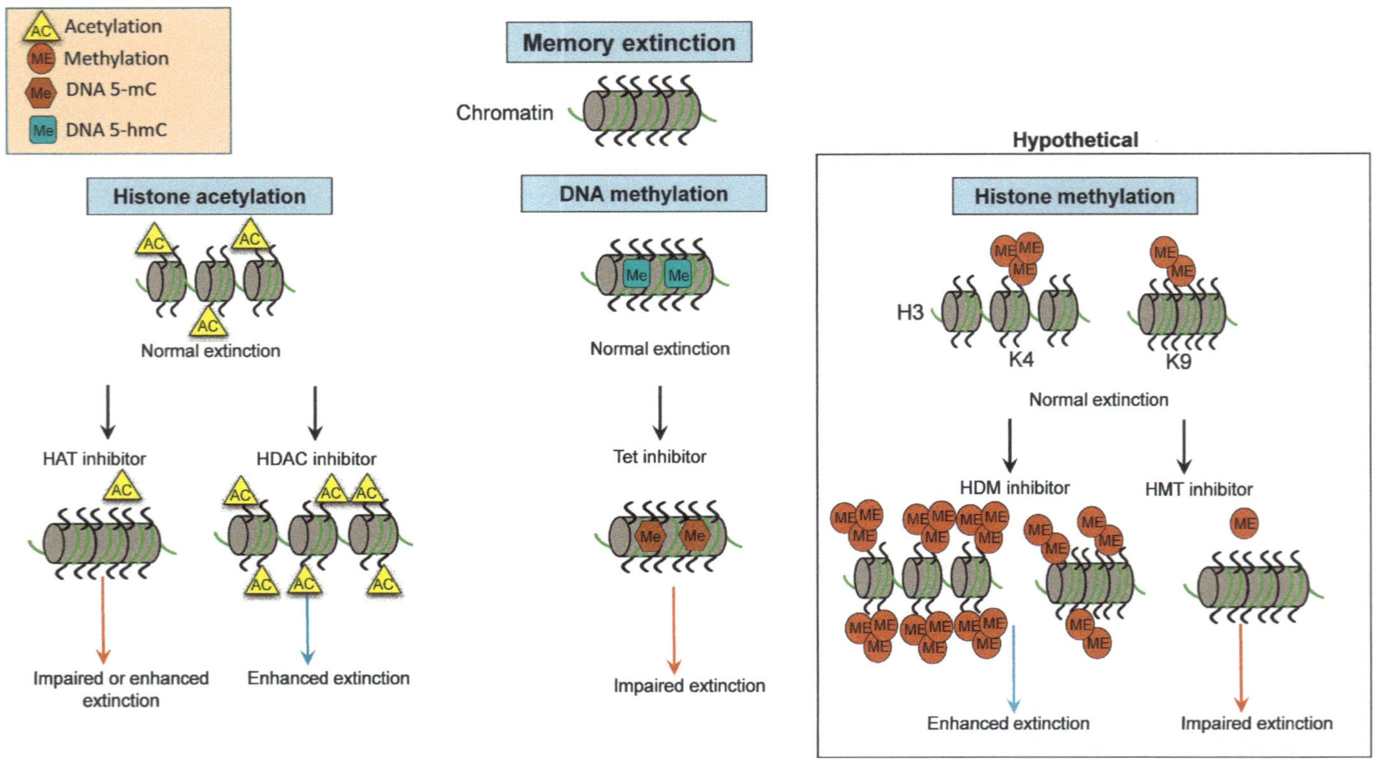

TIMOTHY J. JAROME *ET AL.*, FIGURE 1.5

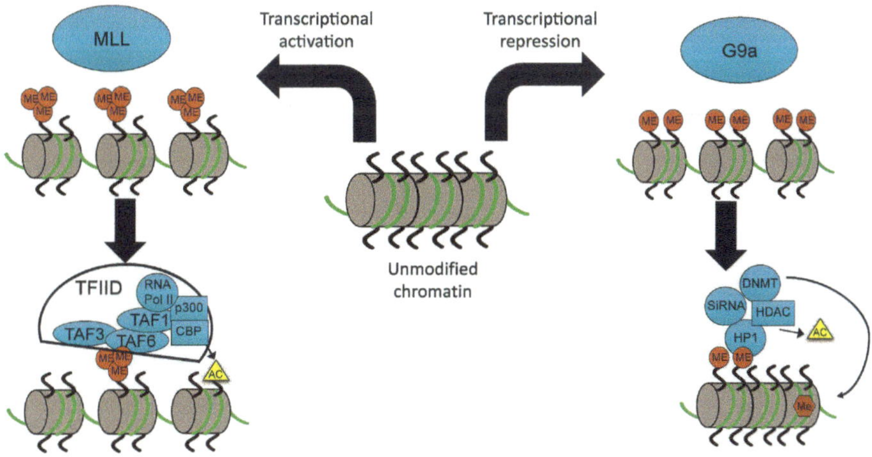

TIMOTHY J. JAROME *ET AL.*, FIGURE 1.6

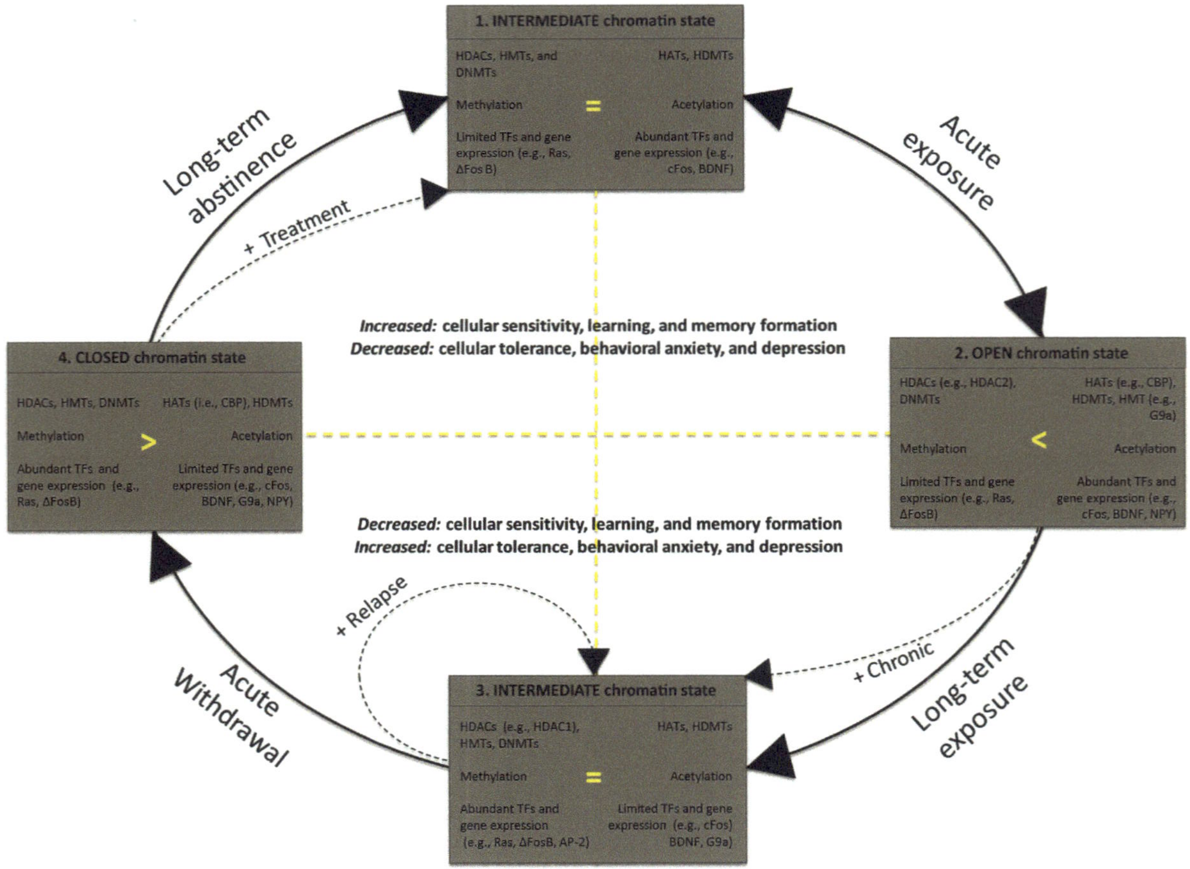

LEAH N. HITCHCOCK AND K. MATTHEW LATTAL, FIGURE 3.1

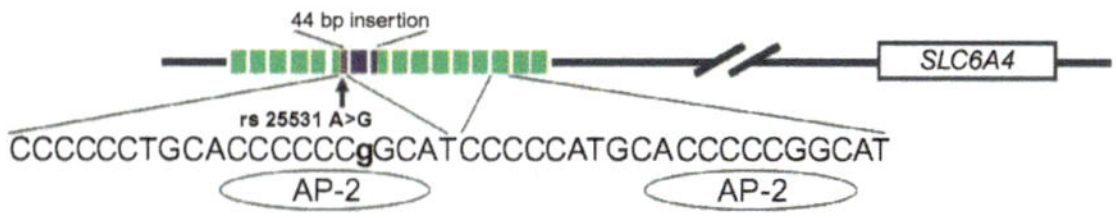

SERMSAK LOLAK *ET AL.*, FIGURE 5.1

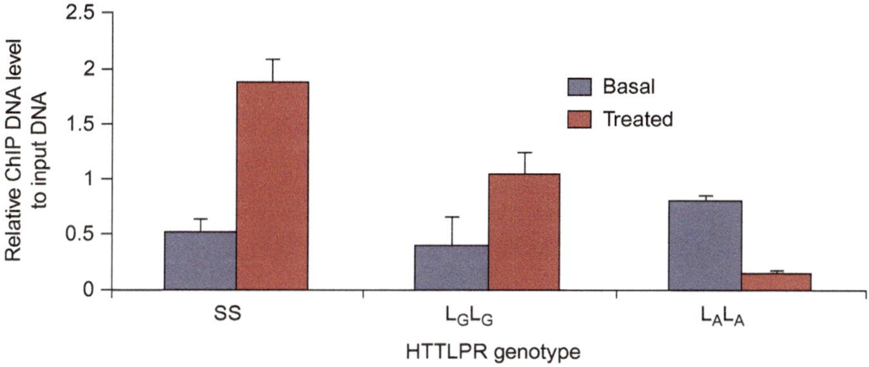

SERMSAK LOLAK *ET AL.*, FIGURE 5.2

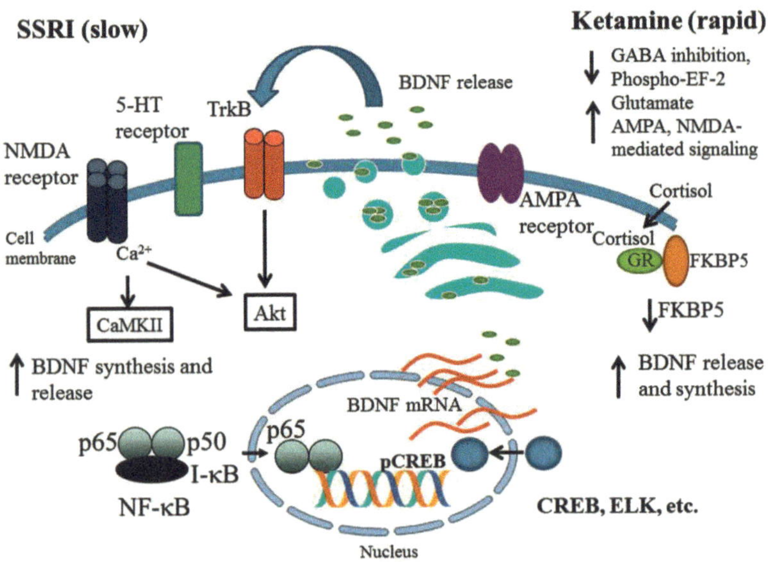

SERMSAK LOLAK *ET AL.*, FIGURE 5.3

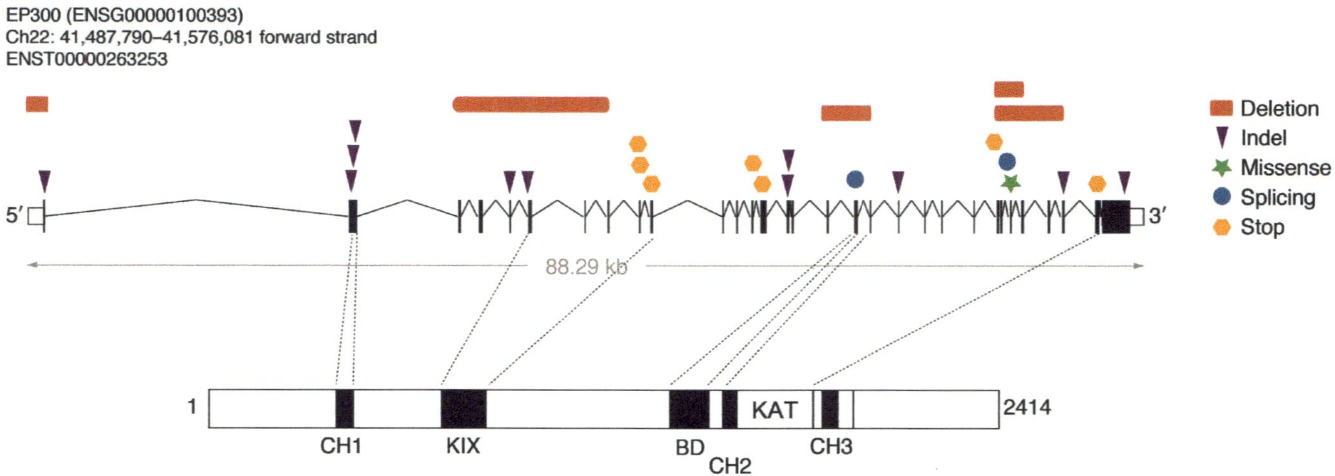

Jose P. Lopez-Atalaya *et al.*, Figure 6.2

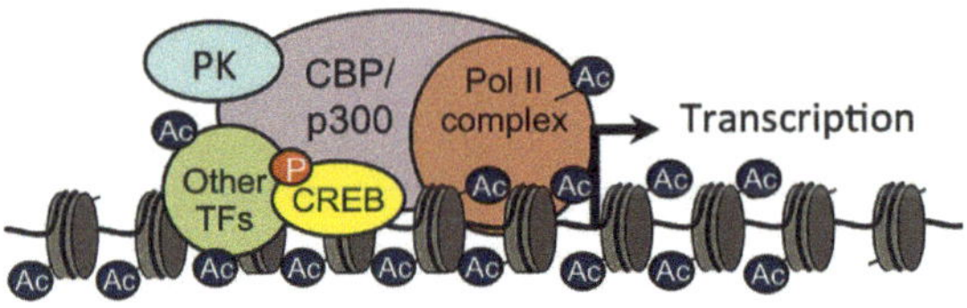

Jose P. Lopez-Atalaya *et al.*, Figure 6.3

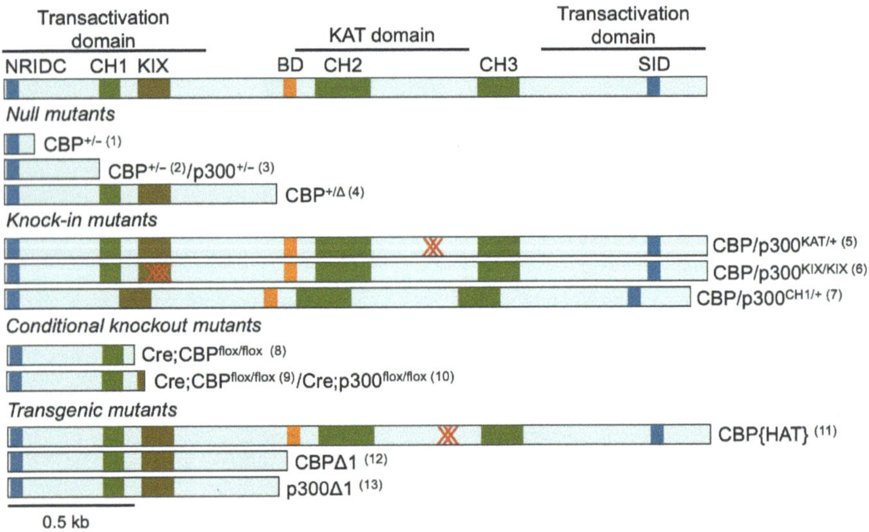

Jose P. Lopez-Atalaya *et al.*, Figure 6.4

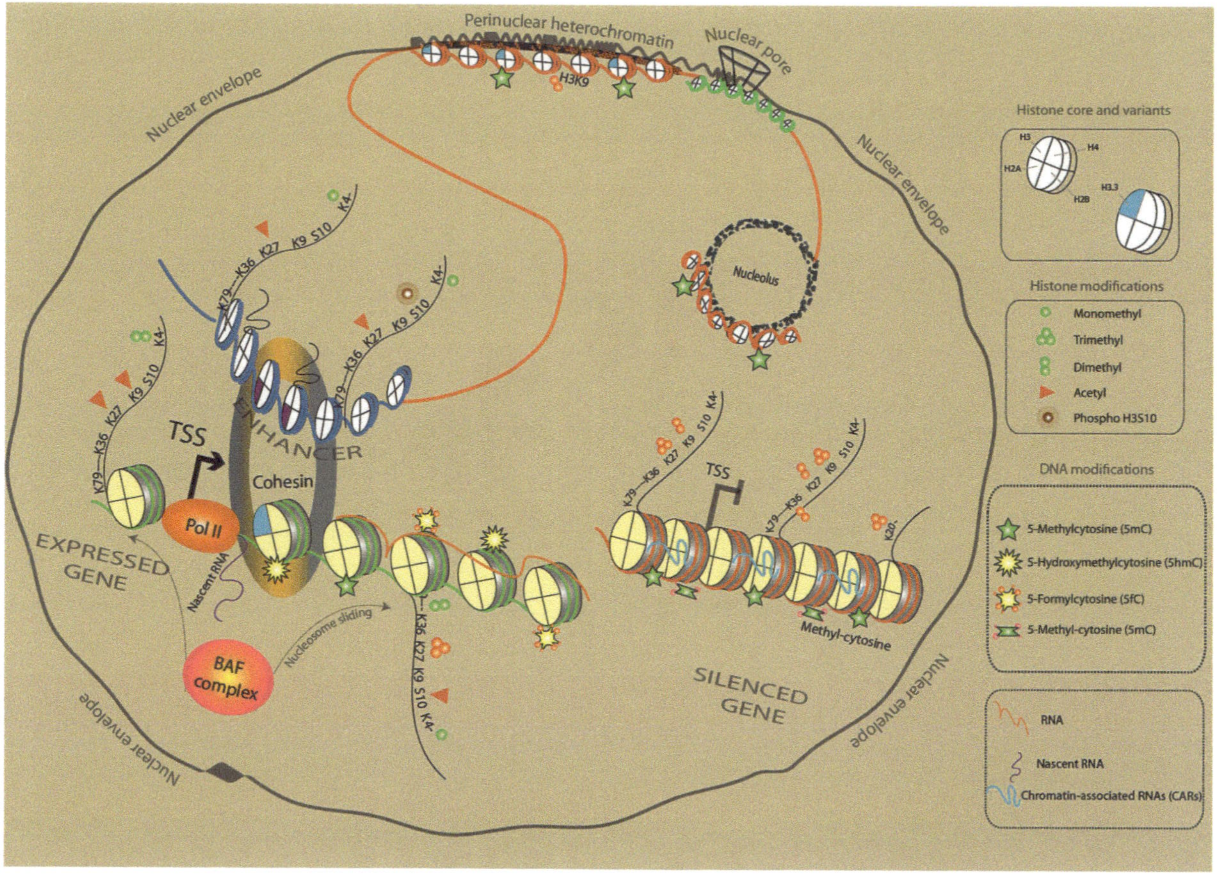

AMANDA MITCHELL *ET AL.*, FIGURE 8.1

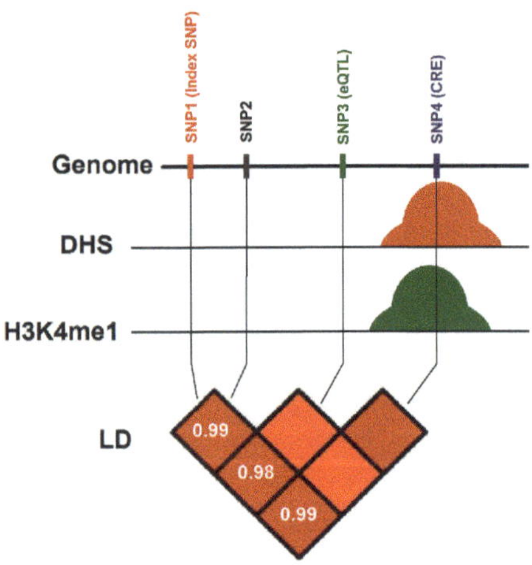

AMANDA MITCHELL *ET AL.*, FIGURE 8.2

FSC
www.fsc.org
MIX
Papier | Fördert gute Waldnutzung
FSC® C083411

Kolibri 360 GmbH
Ferdinand-Jühlke-Straße 7
99095 Erfurt, Deutschland
produktsicherheit@kolibri360.de